Research and Development on the Prairies

A History of the Saskatchewan Research Council

LAURIER L. SCHRAMM

LAURIER L. SCHRAMM

The Author:
Dr. Laurier L. Schramm
Saskatchewan Research Council
125 – 15 Innovation Blvd.
Saskatoon, Canada, S7N 2X8

Print ISBN: 978-0-9958081-3-3
ePub ISBN: 978-0-9958081-1-9

LAURIER L. SCHRAMM

DEDICATION

All proceeds from the sales of this book will go to the Saskatchewan Research Council's Technology in Action Fund - a perpetual memorial fund established to help the people of Saskatchewan develop their province as a highly skilled, fair, desirable and compassionate society with a secure environment through research, development and the transfer of innovative scientific and technological solutions, applications and services.

LAURIER L. SCHRAMM

1947 src 2017

"The need for a scientific organization responsible to the Provincial Government was realized shortly after the second world war. Indications that the economy of the province was entering a phase of rapid change presenting new opportunities and problems in the fields of science and engineering prompted the Provincial Government to establish the Saskatchewan Research Council in 1947. ... the consequent benefit to the provincial economy relative to the expenses of the Council has been beyond expectation."

T.C. "Tommy" Douglas
Premier of Saskatchewan, 1958 [1]

"... the Saskatchewan Research Council has attempted to initiate and encourage research on a number of problems relating to the economic life of the Province and to the improvement of living conditions in the Province ... The impact of this research ... is multiple. It advances knowledge, helps develop our provincial resources, ... helps us poor mortals reach out to the stars, and, ... helps us to do just a few of the impossible things we people in Saskatchewan have always been capable of doing."

Dr. John W.T. Spinks in 1958
shortly before becoming
President of the University of Saskatchewan [2]

"Over the years in Canada an impressive net-work of Research Councils has been built up, ... and they have, I think, constituted a great and well-recognized Canadian achievement ... there certainly seems to be scope for government laboratories in the future ... I very much doubt that there is any danger of the Saskatchewan Research Council running out of problems [to address]."

Dr. E.W.R. "Ned" Steacie
President of the National Research Council, 1958 [3]

LAURIER L. SCHRAMM

CONTENTS

LAURIER L. SCHRAMM

FOREWORD

When I was asked to write this forward to the history of the Research Council my response was ... that will be easy. My background is in research and development. I have been a member of the Board for five years. I knew the Council and its many contributions to research, innovation and the people of Saskatchewan. However, reading the Council's history made me realize I had overestimated my knowledge and failed to appreciate Saskatchewan Research Council (SRC) contributions to the development of the prairies.

Start with the basics. What does the SRC do? It applies science in the support of economic development. Many organizations share this purpose. The National Research Councils in the United States of America and in Canada were founded in the early twentieth century to meet the needs of the modern economy. They work at the national level; SRC is different. It was developed to meet needs of local industry to serve the people of Saskatchewan.

SRC is an industry enabler. It provides technical support during industry development. It makes long-term commitments. For example, in 1994 SRC launched a uranium exploration program with the support of the province and the four uranium companies that were active in Saskatchewan. That program led to the development of a geological model that was adopted by industry and used in their exploration strategy. As a Board Member, I knew our research program supported the industry. What I didn't know was SRC's role in establishing the industry in Saskatchewan.

SRC is a world leader in technology areas important to the Prairie economy. It does minerals. SRC's uranium assay laboratory is the largest in the world. It has one of the largest diamond assay laboratories in the world. SRC is the preferred supplier for geoassay services to De Beers worldwide. Its potash geoassay laboratory is the largest in the world.

SRC does energy research. Their Pipe Flow Technology Centre supports the development of Saskatchewan's oil and gas industry. It is the Western Canadian pipeline research hub. SRC field research focuses on problems related to the production and recovery of Western Canada's petroleum resource. SRC provides the industry with solutions.

SRC does more than support industry. It responds to the needs of the people of Saskatchewan. In 2006 SRC was contracted by the province to manage the remediation of the former Gunner and Lorado uranium mine and mill sites, as well as 35 satellite uranium mines in the Uranium City area of Northern Saskatchewan. All 37 of these sites were abandoned orphans, remnants of the Cold War. The project focused on environmental protection with significant public safety objectives, because of the obvious radiation safety aspects and the hazards related to aging structures, openings to underground shafts and chemical hazards.

The success of Project Cleans, SRC's remediation program, is documented in the book. My comments are meant to address the Corporate Social Responsibility model adopted by SRC. SRC worked in concert with the local community to maximize the local benefits, training and economic, during the many years of the remediation. SRC supports people.

I have provided this snapshot of SRC today to provide context. Dr. Schramm, in Research and Development on the Prairies: A History of the Saskatchewan Research Council, describes the evolution of the Council. This is a distillation of what I learned: Good ideas are resilient. SRC was launched in 1930. The Great Depression swept across North America, wheat prices collapsed, public spending stopped, and SRC was a casualty of the times. When the economy turned SRC returned.

I learned that SRC responds to challenges. It took on new research lines when SRC acquired the program responsibilities of the former Forest First organization in Prince Albert. It embraced environmental remediation when it launched Project Cleans. SRC's new well-test centre focuses on extracting additional value from end-of-life wells was its response to the changing nature of the industry.

Finally, I learned that SRC was a product of the prairies. Saskatchewan farmers are famous for their ability to keep equipment operational. With pliers, wire, and wits they keep complex equipment up and running. No challenge is too daunting. SRC is grounded in this spirit: hard work, perseverance, and a strong entrepreneurial character have made the Research Council successful. The collective benefits to the people of Saskatchewan have been breath-taking.

This book on the history of research on the prairies is not destined to become a best-seller, but for those of us interested in research, innovation, and public service it is a page-turner.

Dr. Dennis Fitzpatrick, Board Chair
Saskatchewan Research Council

PREFACE

In 2017 the Saskatchewan Research Council reached its 70th year of continuous operation. This book is mostly organized into eras, as there have been five eras of distinctly different leadership and strategy. In what I have termed The Early Years, 1947-1955, SRC had a Management Board and was in start-up mode. Following that the Board of Directors became a Governance Board and the primary functions of leadership and management fell to a series of Chief Executive Officers (CEOs) whose job it was to build-up the company (The Building Years, 1956-1972), establish the maturity needed to sustain the business niche and the functional capacity that had been built (The Maturing Years, 1972-1983), test how far the company could stretch into market-pull commercial operations (The Commercial Years, 1983-2000), and then build on what had been achieved in all of the previous eras by setting a balanced stakeholder-needs focus, establishing a core business position within the innovation continuum, and aggressively pursuing growth and impacts within those (The New Millennium Years, 2001-2017).

LAURIER L. SCHRAMM

ACKNOWLEDGMENTS

In 2001 a young-ish (these things are relative) industrial scientist and executive arrived in Saskatoon to become SRC's new President and CEO. SRC had just been through a wrenching crisis that had in part created this opportunity, and as is the case with many such career transitions I arrived with many conflicting thoughts and emotions. I had come for several specific reasons, mostly related to new challenges, new opportunities to learn, and the potential to accomplish something worthwhile. I have been blessed by getting everything I asked for and more, but the reasons I came are different from the reasons that I have remained. I stayed because of the people. The best thing SRC has given me has been the opportunity to work with the finest, most welcoming and engaging, dedicated and hard-working, and intellectually stimulating people it has ever been my privilege to encounter, let alone work with. To all of my friends and colleagues at SRC, and to all of our predecessors at SRC, this book is dedicated to you.

In my journey at SRC, I have been fortunate to work with three great Board Chairs, Keith Hanson, Craig Zawada, and Dr. Dennis Fitzpatrick. Each of them served in differing times and circumstances, but each was the right Board Chair for the times, a strong partner in vision, strategy, and governance, and each was visionary, supportive, and inspiring.

I would like to acknowledge and thank Wanda Nyirfa, Ann Marie Schramm, Ernie Pappas, and Craig Murray for reading drafts, making constructive suggestions and, most importantly, for their support and encouragement.

"If I have seen further [than others]
it is by standing upon the shoulders of giants."

Isaac Newton, in a letter to Robert Hooke on Feb. 5, 1675.
Adapted from Bernard of Chartres, 12th Century philosopher.

LAURIER L. SCHRAMM

1 BEGINNINGS, 1930 - 1947

At the beginning of the 20th-century organized research, development, and technological innovation were quite rare but this was about to change with the advent of industrial research organizations. The United States' National Research Council (U.S. NRC) was formed in 1916 and seems to have quite quickly become interested in the role of research in helping industry advance (hence the term *"industrial research"*). In 1928 Maurice Holland, the first head of Engineering and Industrial Research at the U.S. NRC, published what may have been the first description of the steps involved in the *"cycle of research"* leading to technological innovation and industrial growth [4,5]. Holland was an early champion of systematic research, to which contributions are made by numerous researchers, as opposed to the independent, lone-wolf type of inventor. Systematic research and development (R&D), and the linear model of technological innovation were fairly new concepts at the time, as in 1928 research was still a relatively new tool of industry in North America - two of the first U.S. industrial research laboratories had been established by General Electric (in Schenectady, New York in 1900) and Du Pont (in Wilmington, Delaware in 1903). Although Holland's model later turned out to be an oversimplification, the concepts of a development pathway and a systematic approach were critical to the evolution of the modern approach to technological innovation [6].

Meanwhile, Canada had also formed a research council of national scope in 1916. Originally formed by Order in Council as the Honorary Advisory Council for Scientific and Industrial Research, it was further enshrined by an Act of Parliament in 1917 (Bill 83), and in 1925 it would become better known under its "short title" as the National Research Council (NRC) [7,8]. Leading Canadian industrialists and university presidents had been calling for some kind of national effort to boost scientific research and the British government, which had just created a Department of Scientific and

Industrial Research (in 1915) had asked that each of the Commonwealth countries do the same in order that they might all collaborate [8].

The initial objects of NRC were to promote the application of science to Canada's "*secondary industry¹*," to advise the federal government on scientific matters, and to "*put science and industry together for the benefit of the people of Canada*" [7]. NRC's first offices were located in the West Block of the Parliament Buildings (in 1917), it moved into its first, modest laboratory facilities in 1926, and its first large, dedicated laboratory building² (on Sussex Drive in Ottawa) in 1930-32 [7].

NRC's first major project was studying the briquetting of Saskatchewan lignite to replace imported anthracite coal as a fuel [7,8], a project that SRC would take up early in its history as well (see Chapter 2). From the opening of its first laboratory in 1927, it was determined that NRC would pursue a balance of "pure" and "useful" research³ [7]. NRC focused on scientific and industrial research related to Canada's efforts during World War I, and then again during World War II. Probably the most widely supported of NRC's initial programs was the establishment of a Bureau of Standards, which became part of NRC's mandate in 1924 [8]. The other major programs at NRC during this period (1927 to the 1940s) included chemical and biological warfare agents, explosives, radar and sonar, and atomic (fission) energy. Following World War II NRC's focus almost completely shifted to research that was felt to be in the national public interest but not well provided by either the academic or industrial research sectors [9]. This meant establishing linkages with both academia and industry while continuing to focus on a combination of discovery research and applied research initiatives.

The first provincial research council to be established in Canada was in Alberta (see Appendix 9.1). The province had begun to actively evaluate its mineral resources and their economic development potential, and was sufficiently encouraged by the results that The Advisory Council of Scientific and Industrial Research of Alberta⁴, was established in 1921 "*with a view to ascertaining more definitely the mineral resources ... and the possibilities of their development*" [10,11]. Alberta's research council work was halted between 1933 and 1942 due to the Great Depression but resumed operations in 1942 [12], originally working in co-operation with the University of Alberta

---

¹ That is, industries beyond agriculture and natural resources, for which existing federal government departments already had responsibility.

² NRC's "*central research institute*," in its original, grand Sussex Drive building, has been referred-to as the "*temple of science*" [8].

³ Now more commonly referred-to as "discovery" and "applied" research.

⁴ Later renamed Science and Industrial Research Council of Alberta, Research Council of Alberta, Alberta Research Council, and Alberta Innovates – Technology Futures (AI-TF).

and on the university campus, but eventually moving to its own dedicated facilities in 1956 [13].

The next provincial research council to be created in Canada was the Ontario Research Foundation (ORF, announced in 1927 and established by a provincial Act in 1928) [8,14]. A separate Research Council of Ontario was created in the mid-1940s but was dissolved in 1955 and most of its activities were transferred to ORF [15]. See also Appendix 9.1.

The third provincial research council to be established in Canada was in Saskatchewan. A first attempt had been launched as the Research Council of Saskatchewan (RCS), based in Regina, under *"The Research Council Act, 1930."* Unfortunately, this coincided with the beginning of the "Great Depression" in Canada in which the prairie provinces were particularly strongly affected (mostly due to the collapse in wheat prices), and which was to last until the beginning of World War II in 1939. During this period very little funding of any kind was available. In fact, funds were so severely restricted that almost everything depended on the volunteer efforts of the first Board of Directors (the "Council"): no staff were hired, no payments for travel expenses were made, and only an absolute minimum of meetings were held [16,17]. Nevertheless, these early Board members, assisted by several faculty members of the University of Saskatchewan, were able to conduct preliminary work aimed at the feasibility of producing gas from coal, advancing the clay products industry, producing magnesium from natural salt deposits, and manufacturing "rock wool[5]" from dolomite deposits. Other work was directed at improving the subdivisional survey methods being used at the time in Northern Saskatchewan.

There are very few surviving records from SRC's work during this era, but what does exist suggests that these SRC pioneers must have exhibited similar characteristics to those of the province's pioneers: hard work, perseverance, and a strong entrepreneurial spirit. Despite these efforts, the prevailing economic conditions made it impossible to conduct active applied research and development programs in the early 1930s [16-18]. As a result, all Research Council of Saskatchewan efforts were eventually suspended and the Act was repealed in 1935 [19]. However, this was only the first chapter in the history of Saskatchewan's research council, as will be seen shortly[6].

---

[5] Rock wool is an insulation material made by melting rocks such as dolomite, adding binders, and drawing it into fibres.

[6] See also Appendices 9.1 and 9.2.

The
Research Council
Act

being

Chapter 265 of *The Revised Statutes of Saskatchewan, 1930*
(effective February 1, 1931).

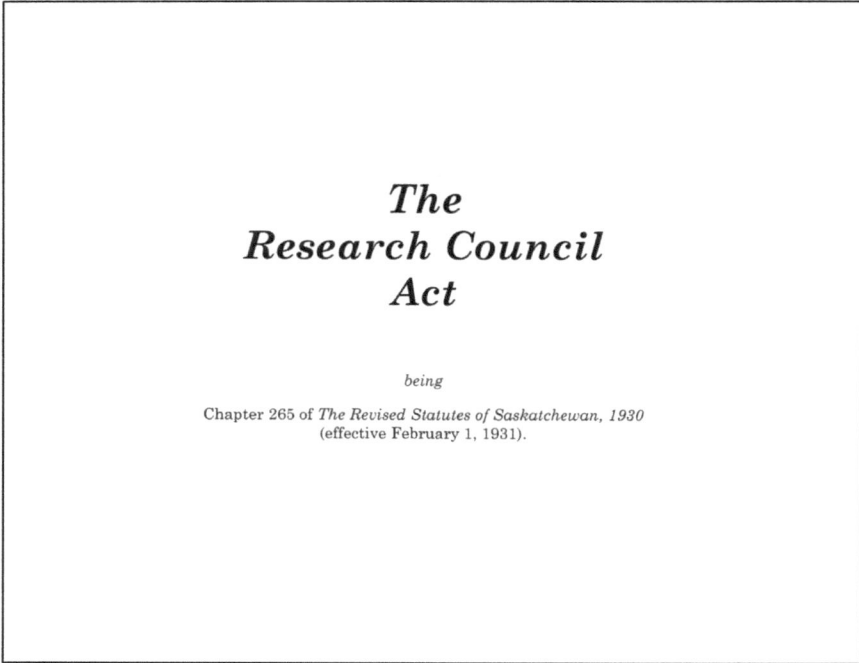

Figure 1.1. *The Research Council Act,* 1930.

The fourth and fifth provincial research councils to be established in Canada (see Appendix 9.1) were the British Columbia Research Council (incorporated as a not-for-profit society in 1944) [20], and the Nova Scotia Research Foundation (NSRFC, established by a provincial Act in 1946) [21]. Subsequently, the New Brunswick Research and Productivity Council was formed in 1962, the Manitoba Research Council in 1963, and the Centre de Recherches Industrielle du Québec in 1969 [22]. Prince Edward Island (PEI) did not create a research council *per se*, however, the PEI Food Technology Centre was incorporated in 1987 by the provincial government as a wholly-owned subsidiary of Innovation PEI, a Provincial Crown Corporation. Its name was changed to Bio|Food|Tech in 2011. In modern terminology, it is a research and technology organization (RTO). Finally, Newfoundland and Labrador created a research council called Research and Development Council formed under its own Act in 2009[7], and later renamed it Research and Development Corp. (RDC).

---

[7] A Newfoundland Research Council Act was actually passed in 1961 but not implemented - a new Act was passed in 2009 to create what is now RDC.

In reviewing the establishment of the provincial research organizations (PROs) in Canada, Le Roy and Dufour conclude that: *"The fact that in some provinces the establishment of the PROs took place much later than in others was not because of any lesser concern for industrial development. Rather, it was a question of when the provincial authorities became convinced that the industrial development of their province could be stimulated through the agency of a provincially sponsored body specially dedicated to this purpose"* [15].

In 1958, Dr. E.W.R. Steacie, then President of the National Research Council concluded in part that: *"Over the years in Canada an impressive net-work of Research Councils has been built up, ... and they have, I think, constituted a great and well-recognized Canadian achievement"* [3].

The Return of the Phoenix. After the end of World War II attention returned to the mineral resource potential and opportunities to grow and diversify Saskatchewan's economy beyond agriculture. As a result, there was a renewal of interest in coordinating research activity in Saskatchewan and *"the concept developed of a research organization acting under the control of the government with the scientific and technical participation of the university"* [18]. The Saskatchewan Government made a successful second attempt to create a research council with the establishment of the Saskatchewan Research Council (SRC), headquartered in Saskatoon, via *"The Research Council Act, 1947."* Under the Act, SRC's primary mission was to *"take under consideration matters pertaining to research and investigation in the field of physical sciences as they affect the economy of the Province of Saskatchewan."* This was broadened in 1978 to *"take under consideration matters pertaining to research, development, design, consultation, innovation and investigation in, and commercialization of, the natural and management sciences, pure and applied, as they affect the welfare of the province"* [23].

The Research Council Act

being

Chapter R-21 of *The Revised Statutes of Saskatchewan, 1978* (effective February 26, 1979) as amended by *The Revised Statutes of Saskatchewan, 1978 (Supplement)*, c.60; and the *Statutes of Saskatchewan*, 1983-84, c.34; 1988-89, c.22; 1991, c.T-1.1; 1994, c.45; 2000, c.23; and 2014, c.E-13.1.

Figure 1.2. *The Research Council Act*, 1947, with subsequent revisions.

2 THE EARLY YEARS, 1947 - 1955

Originally, SRC's head office was situated in Regina and in its first operating year SRC established a Board of Directors[8] and several *"Technical Committees,"* which comprised representatives from government and academia who shared common technical interests [24]. Many of these Members were quite senior and experienced people, including several Deans from academia, several Deputy Ministers from the government, the Deputy Provincial Treasurer, and a Cabinet Minister [24,25]. The Hon. Woodrow S. Lloyd, who was the Minister of Education at the time, was the founding Chair of SRC (see Figure 2.1 and Appendix 9.3). The first meeting of SRC's Council was held on July 15, 1947, and SRC's total revenues for that first year comprised $20,000 [18].

In setting out the initial strategic direction for the company, the principal aim was to undertake applied research targeted at developing *"the resources and economy of Saskatchewan"* but it was recognized that *"it is often necessary to carry on scientific investigations of the basic type before applied problems can be attacked strategically"* [24].

From the beginning [24] SRC anticipated conducting direct research activities at some point in time, but in its early years, SRC simply funded specific applied research projects to be conducted at the University of Saskatchewan[9]. For the most part, the university faculty donated their time and the funds went to the direct operating costs of the projects.

---

[8] At the time this was referred-to as the "Council," and its Members were "Members of the Saskatchewan Research Council."

[9] These university grants started in 1948, were generally in the range $60k to $100k per year, peaked at $118k in 1982/83, and were wound-down to zero by 1989/90.

Figure 2.1 A founding Board member and SRC's first Board Chair was the Hon. Woodrow S. Lloyd (the Minister for Education at the time). He served in these capacities from 1947 through 1960. He became Premier of Saskatchewan in 1961. Saskatchewan Research Council photo [18].

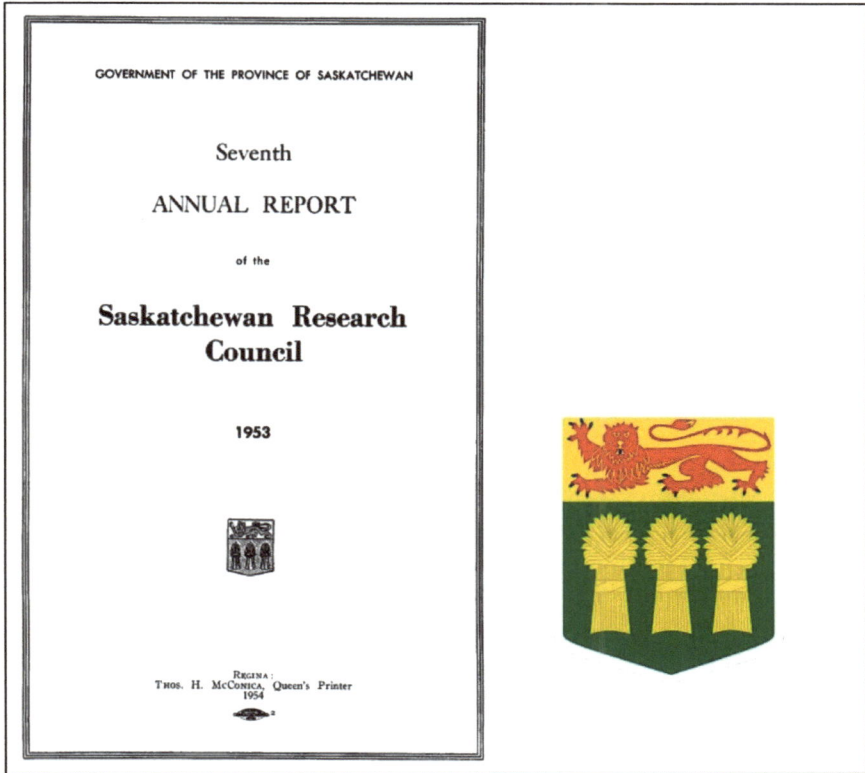

GOVERNMENT OF THE PROVINCE OF SASKATCHEWAN

Seventh

ANNUAL REPORT

of the

Saskatchewan Research Council

1953

REGINA :
THOS. H. McCONICA, Queen's Printer
1954

Figure 2.2 In *The Early Years*, 1947–1955, SRC adopted the Saskatchewan Coat of Arms as its company logo.

The first main applied research projects were aimed at the province's diverse natural resources and, of course, agriculture. This was reflected in projects aimed at such problems or opportunities as [24-33]:

- o Nature and possible industrial uses for wood (i.e. Poplar) lignin and lignite coal (see Figure 2.3),
- o Composition and properties of Saskatchewan clays, including ball (i.e., kaolinitic) clays, and their potential suitability for use in making ceramics,
- o Chemical and physical properties of Saskatchewan volcanic ash deposits (clay-like mineral deposits), and determining the feasibility of using them in commercial products such as concrete,
- o Possible industrial applications for Saskatchewan's sodium sulphate resources, including the manufacture of soluble silicates,

o Uranium exploration, analyses of radioactive mineral deposits in Northern Saskatchewan, and uranium mineral processing research – particularly flotation and ion exchange (see Figure 2.4 below),

o Climate studies, particularly those focused on the conditions responsible for blizzards, causing highway hazards, and Spring rapid thaws, causing floods. In both cases, the two-fold focus was on the meteorology and also on mitigation measures. An example of an early SRC application was the use of this work to guide the design and placement of highway snow fences and windbreaks [27],

o Chemical and economic evaluations of water softeners and detergents for domestic and commercial use (it was noted that Saskatchewan *"has, on average, the hardest water in Canada"* [26], see Figure 2.5 below),

o Petroleum reservoir evaluations, initially focused on the Lloydminster field,

o Improving the performance of Lloydminster Asphalt formulated for road paving,

o Determining ways to improve and sustain cold-weather engine lubrication in vehicles and machinery in order to reduce the winter damage being experienced in one of Canada's coldest winter climates. This involved numerous vehicle engines in a laboratory setting, plus a small fleet of vehicles, and it seems to be the first recorded example of field testing and demonstration [27],

o Building foundation research, particularly into the factors causing seasonal building movements (see Figure 2.6 below),

o A safflower breeding program aimed at creating a better fit with Saskatchewan's growing season and also improved disease resistance,

o Radioactive tracer experiments using P-32, aimed at determining the uptake of various fertilizers by wheat crops,

o Testing and evaluation of moderate-scale, low-head pumps for the irrigation of farmlands in Saskatchewan,

o Determining ways to improve farm animal nutrition, particularly for cattle, sheep, and swine; and including nutritional assessments of frozen wheat, light grains, and forage crops,

o Assessing the most important factors contributing to commercial egg quality, including animal management, nutrition, and genetics,

o Developing frozen food processes, including a frozen, evaporated milk process that was aimed at offsetting winter milk supply shortages (see Figure 2.7 below),

o Investigating the origin of undesirable flavours in commercial fresh-water fish products, and developing means to prevent or remove such flavours,

o   Mosquito studies aimed at reducing or preventing the transmission of mosquito-borne viruses, and

o   Construction of a facility for radio-carbon (C-14) dating of materials in order to assist with archaeological, geological, and even climate change research projects [30,31].

In addition to the above programs, in 1949 SRC assumed responsibility for the Province's contributions to the Prairie Rural Housing Committee (from the Department of Reconstruction and Rehabilitation) [27]. This was the first of several program responsibilities that SRC would assume from the Province and other organizations over the years (see Appendix 9.6).

SRC's first public listing of external publications arising from its work comprised a list of 17 papers and monographs published and several technical presentations spanning 1950 - 1953 [30]. This practice was continued in subsequent years. One of the more significant publications of SRC's "Early Years" was a series of six monographs on the *"Climate of the Prairie Provinces and Northwest Territories"* by University of Saskatchewan professor Dr. B.W. Currie, covering precipitation, temperature, snowfall, ice and soil temperatures, wind and storms, and vegetative and frost-free seasons [31].

SRC's reports of the time show that in order to advance the research work it was often necessary to also develop new and/or improved physical and analytical measurement techniques and new laboratory and field testing equipment (see for example [27,29] and Figure 2.6). The virtually continuous development of new and improved laboratory techniques was to become a characteristic feature of SRC's work that has continued to the present day.

Beginning in 1949 SRC established a small graduate scholarship program at the University of Saskatchewan[10] to encourage students to undertake advanced studies in the natural sciences, and in the hope that this would further SRC's goal of advancing research applied to Saskatchewan's natural resource interests [27]. SRC funded and managed this scholarship program until 1971, at which time it was superseded by a Provincial Government scholarship program [34] (see also Appendix 9.7). For many years afterwards (until 1988) SRC continued to support this program by selecting or advising on the selection of the scholars, and many such scholars worked on SRC projects.

---

[10] This was extended to the University of Regina as well, in 1974, when it became independent of the University of Saskatchewan.

Figure 2.3 Extracting waxes from lignite coals from Southern Saskatchewan (Saskatchewan Research Council, 1952 [28]).

Figure 2.4 Flotation concentration of uranium ores from Northern Saskatchewan (Saskatchewan Research Council, 1952 [28]).

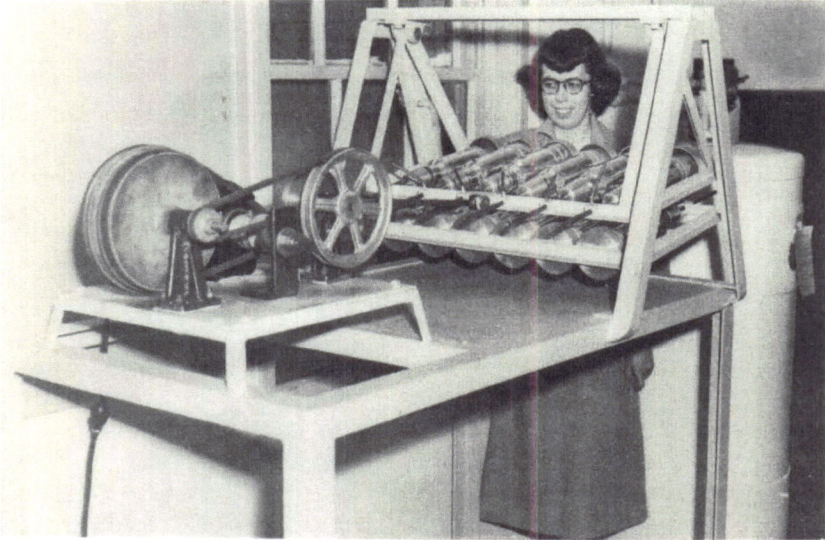

Figure 2.5 Testing the efficiency of detergents in Saskatchewan hard waters (Saskatchewan Research Council, 1952 [28]).

Figure 2.6 SRC neutron moisture meter, for remotely detecting the moisture content of sub-surface clays (Saskatchewan Research Council, 1956 [32]).

Figure 2.7 Developing a process for a frozen, evaporated milk product (Saskatchewan Research Council, 1952 [28]).

In 1950 SRC was beginning to look beyond its internal programs and its activities at the university, and it began to develop relationships with other research organizations in Canada, beginning with the National Research Council of Canada (NRC), the "Forest Products Research Laboratory[11]," and extending to "other provincial research councils" [25]. This was particularly good timing as NRC had just, in 1948, opened a Prairie Regional Laboratory in Saskatoon that was billed as being the "largest single research laboratory in Canada west of the Great Lakes" [8].

The other provincial research councils at that time[12] were the Research Council of Alberta, the Ontario Research Foundation, the British Columbia Research Council, and the Nova Scotia Research Foundation. Such relationships, especially those with NRC and the other provincial research organizations, would vary in scope and intensity from year to year but would ultimately stand the test of time and endure to the present day.

In 1951 it was noted that SRC's research programs were proceeding well

---

[11] This may have been the Forest Products Research Laboratory, Oxford, UK, which was active at the time.

[12] Other provincial research councils, in other provinces, would be formed in later years. See Appendix 9.1.

and that the next phase of SRC's evolution as an organization *"may well be the development of procedures and facilities designed to handle problems in research referred to it by Saskatchewan industry"* [25], in addition to those suggested by government and academia. By 1953 this thinking had extended to the possibility of revising *The Research Council Act* to enable SRC to accept funds from other sources than simply the provincial government, something that was incorporated into the *Act* by a revision in 1954 [30,31].

In developing and launching its earliest programs, SRC's first Core Value was established in terms of a vision that in all of its endeavors SRC would: *"utilize its resources in such a way as to make some real contribution toward fostering and maintaining a live and free spirit of scientific enquiry"* [27]. At the same time, the ultimate goal was to produce useful results that would be adopted into industrial practice so SRC developed its first Patent Policy, based on the practices that had been developed by the National Research Council [25]. In concert with this, the Research Council Act was revised in 1951 to enable SRC to manage patent matters [28]. The Patent Policy is just one example of SRC's operations having become sufficiently varied and complex that by 1951 it had become necessary to begin developing and adopting a range of policies and standard procedures [28].

There was no full-time leader or manager for SRC during *The Early Years* but a part-time position of Coordinator of Research was created in 1949, by which a faculty member from the University of Saskatchewan would be cross-appointed to SRC and devote approximately half-time to the SRC role [18]. The first SRC Coordinator of Research was Dr. Elvins Y. Spencer, a Chemistry Professor at the university, who served from 1949 through 1951 [27,28]. He was succeeded by another Chemistry Professor, Dr. Thorvaldson (see below) who, having "retired" from the university served practically full-time in this capacity from 1951 through 1956 [28].

By 1953 SRC had hired its first full-time employee, to provide administrative support, Mrs. Florie V. Elvin (see Figure 2.9) and opened a Saskatoon office[13]. Mrs. Elvin served for more than 20 years, until 1974/75.

---

[13] This first Saskatoon office was located in the Chemistry Department at the University of Saskatchewan [18].

Figure 2.8. SRC's first part-time Coordinator of Research was University of Saskatchewan Chemistry Professor Dr. Elvins Y. Spencer. He served from 1949 to 1951. Saskatchewan Research Council photo [18].

Figure 2.9 SRC's first full-time employee, Mrs. Florie V. Elvin (Saskatchewan Research Council, 1953).

In 1953 SRC discussed with NRC the possibility of assuming regional responsibility for NRC's Technical Information Service (TIS), a program aimed at assisting industry by making appropriate research information available to it [30]. These discussions were successful and the TIS program at SRC launched in 1954[14]. This not only enabled SRC to engage with and

---

[14] See also Appendix 9.1.

help industry, small companies, and the public across Saskatchewan, but the funding assistance provided by NRC most likely marked the first external revenue to SRC beyond its core provincial investment [18,31]. By 1954 SRC had an office in Saskatoon, as well as Regina and Mr. Ibhar S. Evans, an experienced engineer, was hired to manage the TIS service from the Saskatoon office. SRC's TIS also marked the first substantial collaboration with NRC, thus beginning a collaborative relationship that was to continue to the present day.

At the same time (around 1953) SRC began to develop close working relationships with entities within the provincial government that share an interest in natural resource economic development, particularly the Department of Natural Resources. This marked the beginning of a relationship that would develop and mature to the present day with what is now the Saskatchewan Ministry of Economy.

Looking back, *"The Early Years"* of 1947 through 1955 can be viewed as years in which SRC was solely "shareholder-needs driven" and functioned almost entirely as a strategic project selection and management entity through which its work was done under grants awarded to the University of Saskatchewan. Not everything went smoothly. Reports from this era frequently referred to the difficulties associated with finding and engaging competent research students and graduate assistants, and of the difficulties involved in accelerating urgent or priority projects [18,28,29,31,32]. By 1952 and 1953 consideration was being given to the employment of permanent, qualified staff, particularly if SRC was to continue to work towards practical application of its research programs [18,29,30]. Furthermore, although SRC was based in Regina virtually all of its laboratory-based research was being conducted in laboratories located in Saskatoon. In 1955 the Board urged the provincial government to approve construction of a permanent headquarters, laboratories, and pilot plant facilities, together with the hiring of a team of permanent technical staff that could undertake a full-time program of practical scientific and engineering research, testing, and development [31]. This proposal was received favourably and planning for a substantive research program, a dedicated facility, and a director to lead and manage the company were all conducted in 1955 [32].

To be clear, the early SRC had benefitted tremendously, even critically, from the strong support and assistance of the University of Saskatchewan, its faculty, students, and its facilities. However, it was found that it was *"not possible to get as much investigational work done as [was] needed to keep pace with the increasing level of industrial activity"* in Saskatchewan [33]. In addition, it was felt that SRC would need to take on an increasing number of projects that would be useful, even necessary, to industry but which would have "little academic value," as they would neither further advanced education nor discovery (fundamental) research [33]. In 1956 another revision to *The*

Research Council Act empowered SRC to "employ full-time staff, build a laboratory, and otherwise extend its activities [33].

The converse was also true in that SRC's "University" program had a beneficial effect on the University of Saskatchewan. Shortly before he became President of the University of Saskatchewan, Dr. John W.T. Spinks remarked that some of the impacts of the University program on the University of Saskatchewan, in addition to the financial support, were that it "*... advances knowledge, ... helps train our students,* [and] *stimulates our professors ...*" [2].

In many ways, SRC's "*Early Years*" could also be considered the "*Thorvaldson Years.*" One of the founding Board Members of this era was Dr. Thorbergur ("TT") Thorvaldson, who helped establish SRC while serving as Dean of Graduate Studies at the University of Saskatchewan, and who for many years continued to support SRC after his retirement from the university[15].

The 1949 Annual Report notes that Dr. Thorvaldson had "*continued to devote considerable time and effort to the work of the Council*" and that "*His contributions to the work of the Council and to the welfare of the people of this province are beyond measurement*" [27]. Similar references to Dr. Thorvaldson's contributions to SRC can be found in subsequent years. In addition to serving as a Board Member until 1965 he served on the Technical Committee and, from 1951 to 1956, he served as SRC's second part-time Coordinator of Research[16] [18,25,28-35] (see Figure 2.10). He helped SRC get started and set the stage for SRC's next era of evolution – "*The Building Years*" (see Figure 2.11).

---

[15] Dr. Thorvaldson had an extremely distinguished career at the University of Saskatchewan as well, having served as Professor of Chemistry, then Head of the Chemistry Department, later the first Dean of the College of Graduate Studies, and then Dean Emeritus.

[16] Succeeding Dr. E.Y. Spencer, introduced earlier, who had only been able to serve for approximately one year before a career move took him to Eastern Canada [28].

Figure 2.10 Dr. Thorbergur ("TT") Thorvaldson SRC's second part-time Coordinator of Research was a University of Saskatchewan Chemistry Professor and Dean. Saskatchewan Research Council photo [18].

"The Saskatchewan Research Council hereby recognizes the great contribution Dr. Thorberger Thorvaldson has made to science in Canada and especially his profound influence in the development of the Saskatchewan Research Council. His depth of experience in research, much of which was related to the problems of this Province, naturally resulted in him being one of those primarily responsible for establishing the Council. His work as co-ordinator during the early days of the Council did much to lay a sound basis for the present vigorous organization. As a member of the Council throughout its history and, for a number of years chairman of the Technical Committee, his influence has had a very great and lasting effect on both the program of research and the organization of the Council"

Figure 2.11 Text from a special scroll presented to Dr. Thorvaldson upon his retirement in 1965 [35].

3 THE BUILDING YEARS, 1956 - 1972

In January of 1956 SRC hired Dr. Tom Warren as SRC's first Permanent Head [33]. He served in this capacity, as Director of SRC, until May of 1972 [33,36] (See Appendix 9.4). Dr. Warren led the development of a dedicated headquarters, laboratories, and pilot plant facilities, built-up a team of dedicated scientific, engineering, and other technical staff that could undertake a full-time program of practical research, testing, and development, and he led the building of a more substantial research program. In doing all this he *"guided the Council through its evolution from a granting agency to an independent research establishment with its own laboratories and scientific staff"* [36].

The Industrial Minerals Research Branch of the Saskatchewan Department of Mineral Resources was transferred to SRC in 1956[17], its objectives were to inventory the province's industrial minerals, assess these minerals and determine the best uses for them, and to raise the awareness of these minerals and their potential with industry and the public [33]. From the beginning this involved both laboratory and field work, as the Branch investigated such pure and mixed minerals as: lightweight aggregate (for concrete), marl (as a substitute for limestone), clays (for ceramics), sodium sulphate (probably for the manufacture of detergents and wood pulp), silica sand (for the manufacture of glass and petroleum industry fracturing fluids), lignite (for fuel), and potash (for fertilizer).

---

[17] See also Appendix 9.6.

Between 1956 and 1958 SRC's programs were increasingly grouped by technical specialty into what would become its Divisions of Chemistry, Physics, Geology, Engineering, and Information Services[18] [33,37,38]. The Information Services Division included the Technical Information Service (TIS) program operated in conjunction with NRC (see Chapter 2), and whose services to industry rapidly expanded in 1956. In 1958 the Industrial Minerals Research Branch was re-distributed among several of these new divisions and ceased to exist as a separate organizational unit [38].

Figure 3.1 SRC's first full-time Director, Dr. Tom Warren took office in January of 1956. Saskatchewan Research Council photo [36].

---

[18] In 1958 the Industrial Minerals Research Branch was re-distributed among several of the new divisions and ceased to exist as a separate organizational unit.

The number of SRC employees had risen to 15 by 1956, 19 by 1958, and a further doubling was being contemplated. This turned out to be an underestimate as the staff size would actually grow to 60 by 1968 [39] and to 98 by 1972 [18,36].

The Industrial Minerals Research Branch acquisition brought SRC its first laboratory and its first major field equipment (Figures 3.3 and 3.4). The first pilot plant, involving a rotary kiln, was constructed in 1955 and put into operation in 1956. This pilot plant was used to test lightweight aggregate materials for making concrete [33] (Figure 3.5).

It was decided that the new laboratory building should be constructed in Saskatoon, land for this purpose was leased from the University of Saskatchewan, and design and construction were carried out between 1956 and 1958 [18,37]. Even at this early stage of evolution, it was anticipated that SRC's work would require periodic adaptations to work aimed at *"quick solution of a large number of diversified ... problems ... rather than to exhaustive investigation in a few major fields"* [37], so care was taken to ensure versatility of equipment and facilities. The building was designed to include laboratories, high-head pilot plant space, offices, meeting rooms, shops, map room, and a library (see Figures 3.6 and 3.7). As demands for industry-driven R&D increased, by 1962 it had already become necessary to design a third floor, which was constructed in 1963 (Figure 3.8) [40,41] and enabled a rapidly expanding analytical chemistry group (Figure 3.9) [42].

Late in 1966 a separate building on Quebec Avenue in Saskatoon was acquired for contract research projects [18] (Figure 3.10). This was used to house the large equipment needed for the slurry pipeline program (see below).

A new research council building would be incomplete without instruments and equipment, and over the next few years some very sophisticated equipment, for that period of time, was acquired. Examples include a Librascope General Purpose LGP-30 electronic digital computer, one of only about four in Canada at the time, which was shared with NRC and the University of Saskatchewan (see Figure 3.11), one of the first commercial infra-red spectrophotometers to come on the market, in 1960 (see Figure 3.12), and an electron microscope, one of only a few commercial units in Canada at the time (Figure 3.13).

With growing staff levels, facilities, equipment, and activities came a need to pay attention to workplace health and safety. SRC's safety program was initiated in 1959 with the introduction of a First Aid log book so that safety incidents could be tracked and to enable learning from previous incidents (Figure 3.14).

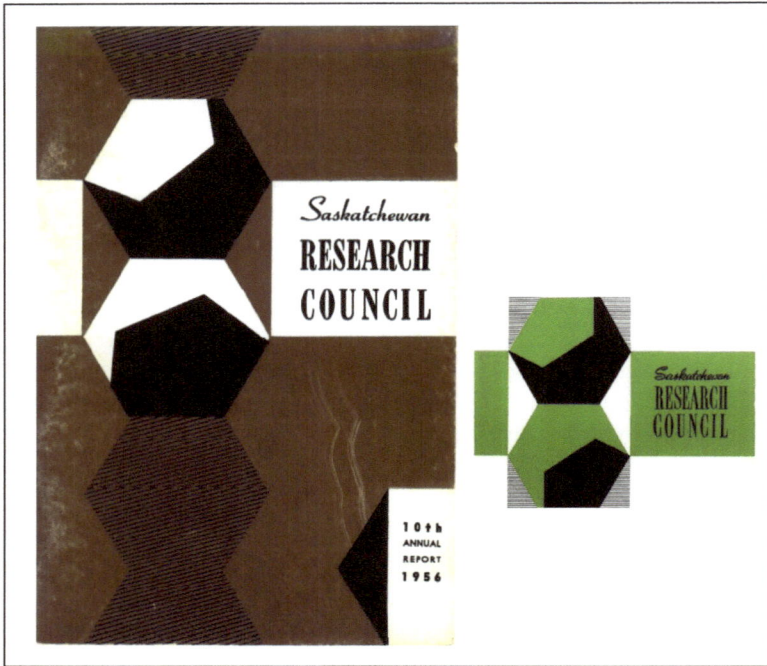

Figure 3.2 In 1956 SRC adopted its first unique company logo.

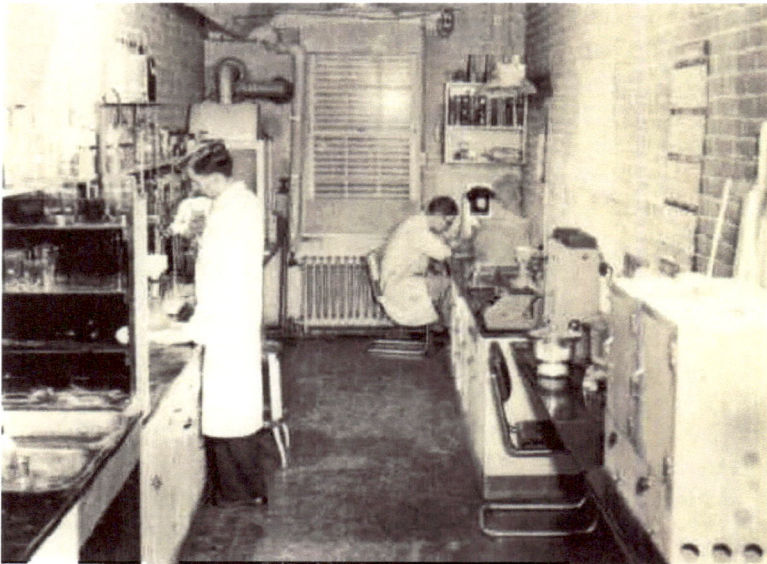

Figure 3.3. SRC's first laboratory for mineral characterization and processing R&D, 1956 [33].

Figure 3.4. Field drilling for mineral core samples in 1956 [33].

Figure 3.5. Rotary kiln pilot plant for lightweight aggregate in cement testing, 1956 [33].

Figure 3.6. Photographs of the new SRC building in 1958; North view (upper) and East view (lower) [38].

Figure 3.7. Plaque from the official opening of the new SRC building in 1958 [43].

Figure 3.8. Photograph of the SRC building with the third-floor addition, in 1963; South view [41].

As SRC's testing and analysis capabilities increased so did requests for technical assistance from individuals, municipalities, industry, and other government departments and Crown corporations, so that as early as 1959 SRC was conducting significant work for Saskatchewan Power Corp.[19] and Saskatchewan Minerals Corp. [44]. SRC's early policy was to conduct such work on demand whenever such services were considered likely to benefit the economy of the province of Saskatchewan or where they were not available commercially [44].

Finally, the Information Services Division, mentioned above, continued to grow throughout *"The Building Years"* [46]. The amount of technical information that could be made available to SRC staff and to industry continued to grow (see Figure 3.16), as did the demands for technical assistance (answering about 1,000 queries per year from industry by 1972), and also outreach activities and publications.

SRC launched several periodical publications including the quarterly *"Scientific and Technical Digests,"* which provided updates on scientific and technical advances that could be of commercial interest from 1962 through 1968. The two longest-standing SRC periodicals were:

o *"The Catalyst,"* a monthly publication[20] aimed at promoting

---

[19] Particularly work related to water management at the Boundary Dam reservoir, whose cooling-basin capacity determined the ultimate generation capacity for the province's coal-fired power utility.

SRC's services, providing information on technological advances, and the importance of applying new knowledge to industrial problems and opportunities, and which was published from 1959 through 1976 with a circulation list of about 5,000 (Figure 3.17).

o *"Industrial Business Management,"* a monthly industrial engineering bulletin[21], which was published from 1964 through 1981 with a circulation list of about 3,000 (Figure 3.18).

Figure 3.9. Illustration of the floor plan for the ground floor of the 1958 SRC building [37].

---

[20] Originally titled *"The Suggestion Box"* when it was launched in 1959. Renamed *"The Catalyst"* in 1969.

[21] Originally titled *"Why?,"* renamed *"Industrial Management Science"* in 1969, and renamed again to *"Industrial Business Management"* in 1972.

Figure 3.10. Photograph of SRC's Quebec Avenue Engineering Laboratories, circa. 1967.

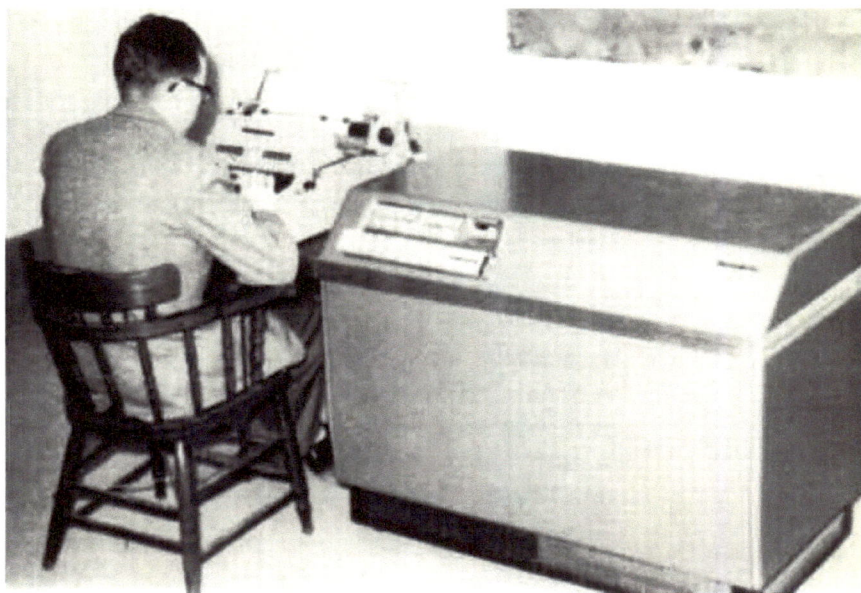

Figure 3.11. The LGP-30 "state-of-the-art" electronic digital computer in 1958 [38].

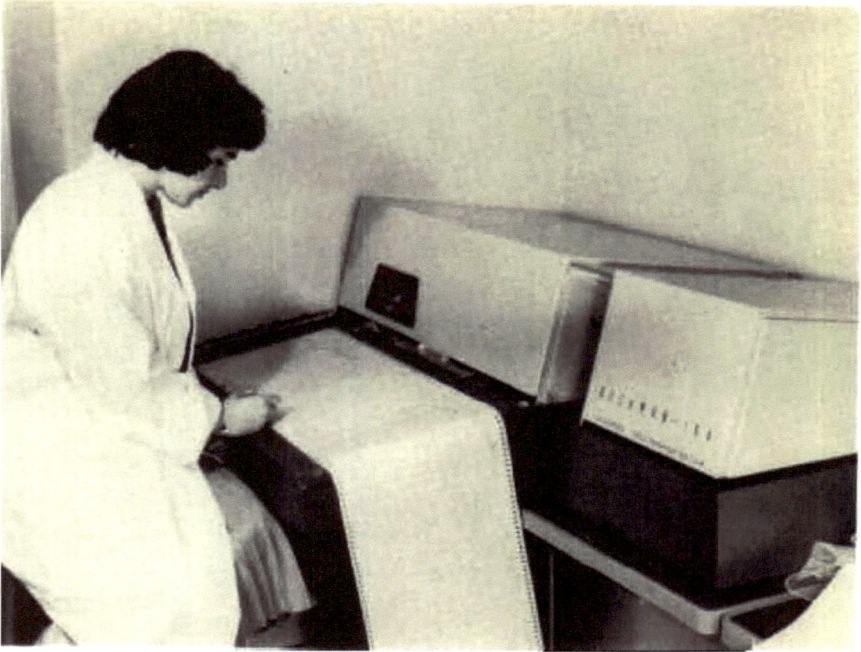

Figure 3.12. One of the first commercial infra-red spectrometers, in 1960 [43].

By 1971 the first commercial book by an SRC staff member had been published[22]. *Physical Environment of Saskatoon, Canada*, by Dr. Earl A. Christiansen (Ed.) brings together the geological, hydrological, and meteorological knowledge of the time for a 1,250 square-mile area centred on Saskatoon, and provides a four-dimensional matrix. The primary intended audience for this book is city planners [50].

---

[22] A list of technical books published by SRC staff members is provided in Appendix 9.8.

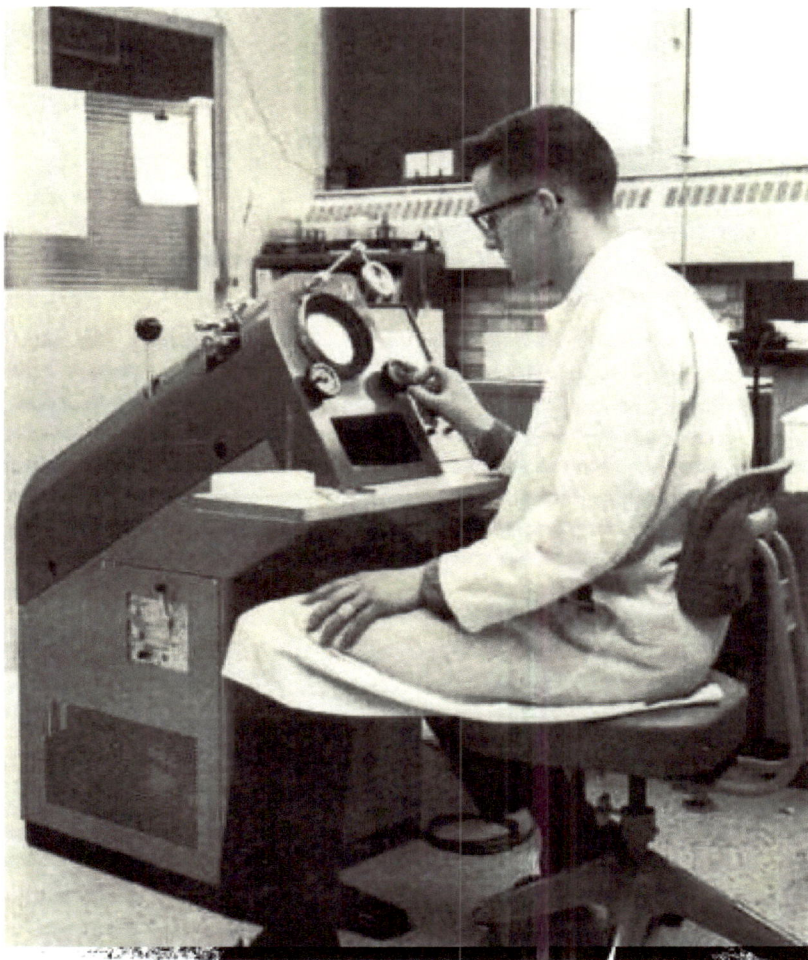

Figure 3.13. Using an electron microscope in 1964 [42].

Figure 3.14. SRC's safety program was launched in 1959.

Figure 3.15. SRC's analytical chemistry lab in 1966 [45].

Figure 3.16. The library in 1965 [35]. By 1972 the library held over 5,000 books and over 10,000 reports, carried over 400 periodicals, and had 25 abstracting and indexing services [36].

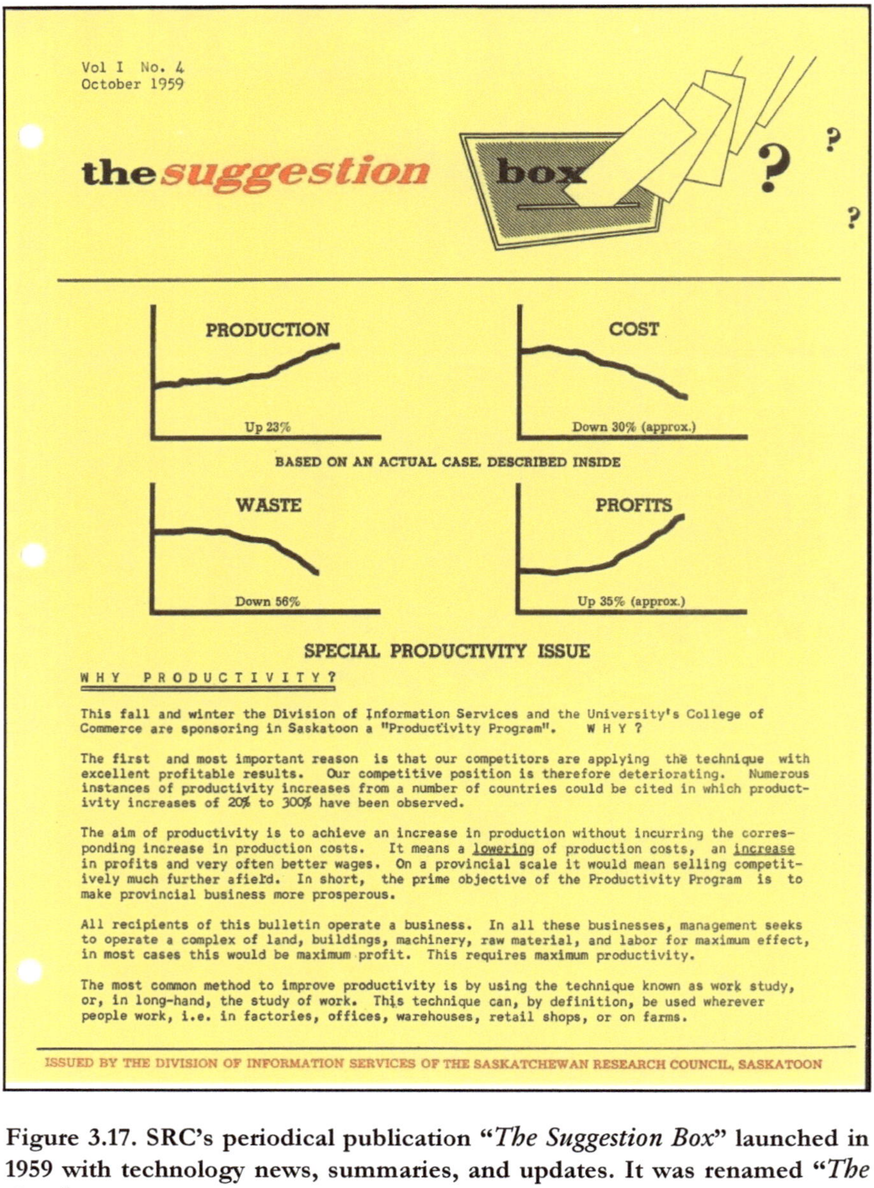

Vol I No. 4
October 1959

the suggestion box ? ? ?

PRODUCTION — Up 23%

COST — Down 30% (approx.)

BASED ON AN ACTUAL CASE. DESCRIBED INSIDE

WASTE — Down 56%

PROFITS — Up 35% (approx.)

SPECIAL PRODUCTIVITY ISSUE

W H Y P R O D U C T I V I T Y ?

This fall and winter the Division of Information Services and the University's College of Commerce are sponsoring in Saskatoon a "Productivity Program". W H Y ?

The first and most important reason is that our competitors are applying the technique with excellent profitable results. Our competitive position is therefore deteriorating. Numerous instances of productivity increases from a number of countries could be cited in which productivity increases of 20% to 300% have been observed.

The aim of productivity is to achieve an increase in production without incurring the corresponding increase in production costs. It means a lowering of production costs, an increase in profits and very often better wages. On a provincial scale it would mean selling competitively much further afield. In short, the prime objective of the Productivity Program is to make provincial business more prosperous.

All recipients of this bulletin operate a business. In all these businesses, management seeks to operate a complex of land, buildings, machinery, raw material, and labor for maximum effect, in most cases this would be maximum profit. This requires maximum productivity.

The most common method to improve productivity is by using the technique known as work study, or, in long-hand, the study of work. This technique can, by definition, be used wherever people work, i.e. in factories, offices, warehouses, retail shops, or on farms.

ISSUED BY THE DIVISION OF INFORMATION SERVICES OF THE SASKATCHEWAN RESEARCH COUNCIL, SASKATOON

Figure 3.17. SRC's periodical publication "*The Suggestion Box*" launched in 1959 with technology news, summaries, and updates. It was renamed "*The Catalyst*" in 1969.

INDUSTRIAL

MANAGEMENT

SCIENCE

Duplicate

INDUSTRIAL SERVICES DIVISION · SASKATCHEWAN RESEARCH COUNCIL

AN INTRODUCTION TO

SYSTEMS DESIGN

by ALAN SCHARF, P.Eng., Industrial Engineer

WHAT IS A SYSTEM?

A system exists where some purpose or function is achieved by the purposeful conversion of some input (resources, materials, information) to some output or means of satisfaction by the application of man and machine skill in a given environment. A full system description will include a description of its physical, or "real life" dimension; its rate dimension or intended direction and rate of change; its control dimension or means of ensuring achievement of norms; and its state dimension or continuing state of existence.

A system can exist in any one of three conditions. A system may exist in a satisfactory condition, an unsatisfactory condition, or in a condition of non-existence. Thus, we may wish to improve or better the operation of a system in a state of satisfactory existence, we may wish to correct a system in a state of unsatisfactory existence, or we may simply wish to design a new system. It can be shown that this third approach, the design of a new system, is also often the best approach when faced with an improvement or correction situation. It is the approach identified as systems design, work design, or the ideals concept.

THE SYSTEMS DESIGN STRATEGY

The systems design approach asks, "Where are we going?" as compared with conventional improvement approaches, which ask, "Where are we coming from?" The strategy or steps in systems design are as follows:

1. DETERMINE THE NECESSARY FUNCTION OR PURPOSE OF THE SYSTEM TO BE DESIGNED.

Function is a statement of what we wish to achieve, not how. Function is not the same as output. The function of an x-ray department may be to assist in diagnosis. The output may be x-ray plates.

Consider the manufacture of an egg carton. If asked what is the function of your company, you may say "to produce egg cartons". Actually you have defined the output of the system, not the function. Why do you want to make egg cartons? To *provide a means of packaging eggs*. That is the function. Why do you want to package eggs? To provide a means of protecting them. That is the function of a larger system which includes other means than packaging for protection. A whole hierarchy of system functions can be established as follows:

UNIVERSITY CAMPUS, SASKATOON – 343-2674 – TELEX SARECO 034-2484

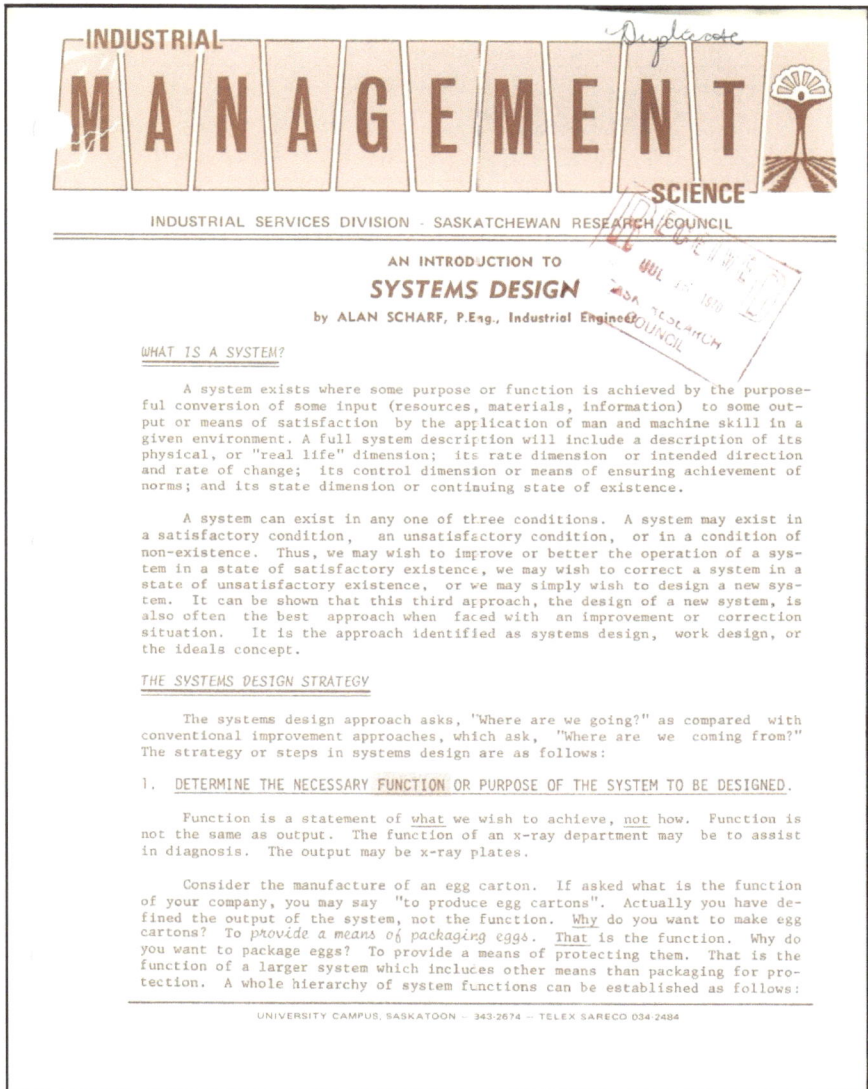

Figure 3.18. SRC's periodical publication *"Industrial Management Science"* launched under the title *"Why?"* in 1964 with industrial management advice and updates. It became *"Industrial Management Science"* in 1969, and was renamed *"Industrial Business Management"* in 1972.

In addition to the range of projects launched and conducted in "The Early Years" (Chapter 2) increased emphasis was placed on projects related to Saskatchewan's oil and mineral resources, such that a more balanced program emerged focused on:

o Agriculture, mostly plant characterization, alternative crop development (such as safflower), irrigation and seed cleaning,

o Water, mostly focused on treating brackish surface water (including a freeze-desalination process) and exploring for groundwater sources for individual, municipal, agricultural, and industrial use. In the early 1960s, this was the focus of about one-third of SRC's work,

o Fisheries, including buffalo fish and fresh-water shrimp characterization,

o Livestock, especially regarding animal nutrition and alternative feeds,

o Manufacturing, principally focused on engine oil additives and lubricating greases (from vegetable oils), ceramic clays, salt cake and other products from sodium sulphate, food products and food product preservation, and forest products,

o Minerals, principally lignite characterization and processing, carbon-14 dating, uranium ore processing, potash resource characterization, and potash solution mining and flotation process development, and copper, iron, and helium deposit characterizations, In the early 1960s, this was the focus of about one-third of SRC's work,

o Petroleum, with initial focus on reservoir characterization, fracturing sand, and natural gas production,

o Construction, principally focused on foundation and highway construction problems, such as settling,

o Transportation, with early and particularly far-sighted consideration being given to the pipeline transportation of solids, and

o Environment, beginning with weather physics in 1960.

As was the case during "The Early Years," there was a continuing need for SRC employees to develop new analytical measurement techniques. For example:

o X-ray fluorescence techniques for the determination of uranium in process samples were developed, which not only greatly reduced analysis times compared with the older fluorimetric method but could be adapted to continuous monitoring in commercial uranium mills [38].

o SRC's advances in equipment and techniques for carbon-14 dating of late-glacial and post-glacial geological materials (see Figure 3.19) almost immediately led to demands for work from organizations across Canada[23] [38]. This led to conducting work for numerous clients and was also supported by Atomic Energy of Canada Ltd.

o In order to effectively and efficiently map buried river channels as part of the groundwater program, SRC staff developed a direct-current resistivity unit capable of making electric drilling measurements to 150 m (500 feet) [44] (see Figure 3.20).

o The need for so much specialized, custom instrumentation and measurement techniques led to the establishment of an electronics laboratory in 1959 and later an instrumentation laboratory (see Figures 3.21, 3.22).

It was also in this era that SRC began working with Canadian oil & gas and mining & mineral companies on projects that were of strategic interest not only to Saskatchewan but other parts of Canada as well. Probably the two most important forerunners of long-standing and high-impact program areas were:

o Working with uranium mining companies (such as Gunnar Mines Ltd., Lorado Mine Ltd., and Columbia Metals Exploration Co.) to help them develop and improve processes for separating uranium from uranium ores (Figures 3.27, 3.29, 3.30), particularly for the Gunnar and Lorado mine and mill operations in Northern Saskatchewan[24] (Figure 3.31) [38]. Such uranium separation process development was scaled-up and became the first pilot plant project in the new 1958 building.

o Working with the Lloydminster Petroleum Association, the Alberta Research Council (ARC), and Husky Oil Ltd. on ways to improve oil recovery from the Lloydminster heavy oil reservoirs[25] [38].

---

[23] Thus, SRC's first ex-Saskatchewan work was initiated, in response to external demand, in 1958 [37].

[24] Approximately 50 years later SRC would come back to the Gunnar and Lorado mine and mill sites to manage their decommissioning and/or environmental remediation. See Chapter 6.

[25] This seems to be the first of many collaborations with the Alberta Research Council, that would continue to the present day.

Figure 3.19. Sample-preparation apparatus for carbon-14 dating of materials (1958) [38].

Figure 3.20. Resistivity apparatus for groundwater mapping (1959) [44].

Figure 3.21. Electronics laboratory (1959) [44].

Figure 3.22. Instrumentation laboratory (1965) [35].

Figure 3.23. White mice used for investigations into feeds and nutrition (1957) [37].

Figure 3.24. Prospecting for groundwater in 1960 [43].

Figure 3.25. A panel taste-testing foods preserved by freezing (1958) [38].

Figure 3.26. Engine testing of lubricating oils with Canola-based additives in 1956 [33].

Figure 3.27. Part of a 15-tonne sample of uranium ore obtained from Higginson Lake (Anglo Barrington Mines Ltd.) in 1956 [33].

Figure 3.28. Winter drilling was used to identify geological features underneath lakes. Here a drill and an SRC-built geophysical logging unit explore beneath frozen Crater Lake near Yorkton [46].

Figure 3.29. Ball mill (left-hand side) and mineral flotation cell (right-hand side) used to study acid leaching processes for uranium ores (1956) [33].

Figure 3.30. SRC's uranium processing pilot plant in 1958 [38].

Three other programs were initiated in 1962 that would later evolve into *"signature platforms"* for SRC:

Groundwater and Water Treatment Studies. With about seventy percent of Saskatchewan's population being wholly or partially dependent on groundwater supplies, and with eighty percent of the resource completed in drift aquifers[26], ongoing research was necessary [54]. What started out as a major effort to locate and map groundwater resources was gradually expanded to include water quality analyses and mapping, and underground water movements [40,42]. Another aspect of this work involved developing methods to treat (desalinate) brackish water [47,48]. Several techniques were developed and pilot-tested, including the highly visible spray-freezing method which created 50 foot-high "ice-mountains" (see Figure 3.33) [48,49].

---

[26] Drift aquifers lie between the ground and bedrock surfaces.

Pipeflow Technology Development. It had been recognized that many strategic materials, notably water, oil and natural gas, were already being transported over long distances by pipeline in Saskatchewan and others, including grain, coal, potash, and sodium sulphate, that required long-distance transportation and might benefit from pipeline transportation methods [40]. Although the science and engineering of liquid-only, and gas-only, flow in pipelines was quite well established the state of knowledge for dispersions, particularly dispersions of solids in a liquid, was insufficient to permit pipeline design. To address this SRC launched a slurry transport program in cooperation with the University of Saskatchewan (U of S) and directed by Dr. Cliff A. Shook from the U of S Department of Chemistry and Chemical Engineering. By 1969 SRC had one of the most elaborate slurry pipeline facilities in the world, with a range of fully instrumented and equipped pipeline loops ranging from 5 to 36 cm (2" to 14") in diameter [46,47] (Figure 3.34). The pipeflow centre would continue to evolve until the present day (see Chapters 4-6).

Climate Reference Studies. SRC's climatology research began in about 1962 with the aim of developing improved weather forecasting and even control, in order to aid the agriculture sector. The University of Saskatchewan had maintained a climatological station for over 40 years. When this was about to be shut-down, in order that the climate record be continued SRC offered to manage the operation of a relocated *"Climatological Reference Station"* with more elaborate monitoring instruments, in cooperation with the university and the Meteorological Service [40,41] (see also Appendix 9.6). The SRC site would also be used to field-test and demonstrate new and emerging meteorological instruments [35]. The SRC *Climatological Reference Station* began observations in October of 1963 (Figure 3.35) and was the best equipped meteorological monitoring station in Western Canada by 1966 [45]. In addition to acquiring the university's historical climate data, SRC also acquired the historical climate records from other Saskatchewan weather stations so that by 1972 SRC had already acquired 70 years-worth of climate observations, i.e., dating back to 1902, in more than 1.5 million records (see Figure 3.36) [36]. Having such a long record of climatological records enables SRC to provide comparisons with "normals" for any given time of year.

Figure 3.31. Gunnar (upper) and Lorado (lower) uranium mine and mill sites in 1958 [38].

Figure 3.32. Pilot plant for soda ash processing (1960) [43].

Figure 3.33. A spray-freezing desalination pilot test in winter 1970/71 [48].

Figure 3.34. The pipeflow pilot plant in 1970 [50].

Figure 3.35. The new SRC *Climatological Reference Station* in 1963 [41].

Figure 3.36. By 1972 SRC had already acquired 70 years-worth of climate observations, i.e., dating back to 1902 [36].

Finally, one of the more unusual projects of *The Building Years* involved the use of hovercraft, purchased in 1971, to access the combination of swamps, shallow, and deep waters needed to conduct hydrological studies (see Figure 3.37). Although this particular application did not develop into a huge success, the program did produce a substantial body of operating information over a wide range of operating conditions, which was contributed to the National Research Council for use in air cushion vehicle developments and standards [36,60].

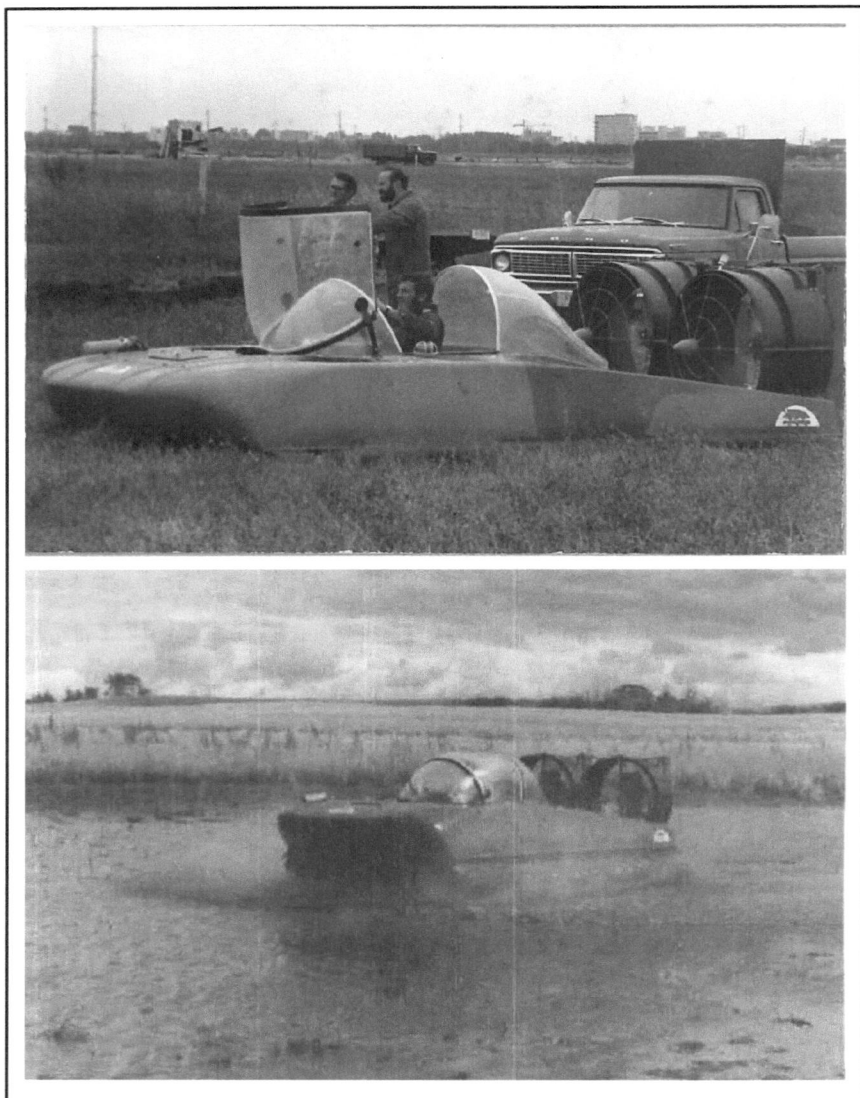

Figure 3.37. Using a hovercraft to conduct hydrological studies in 1971. Top photo Saskatchewan Research Council; lower photo Reference [48]).

It was noted at the beginning of this chapter that as SRC's leader during *The Building Years* Dr. Tom Warren *"guided the Council through its evolution from a granting agency to an independent research establishment with its own laboratories and scientific staff"* [36]. Driven by a focus on its shareholder's needs SRC developed its own dedicated people, facilities, and external funding sources. The combination of demonstrating an ability to deliver practical results of strategic interest to the province and contributing to Saskatchewan's economic growth enabled significant funding increases, which in "The Building Years" averaged 11% per year [18]. With so much new work on practical industry problems the revenues from contract work grew and became a more important part of the revenue mix. The amount of provincial investment increased by an order of magnitude (see Table 3.1) while the amount of contract revenue increased from being insignificant to over half a million dollars, and about half of the provincial investment level.

Although the amount of money allocated to sponsoring applied research at the University of Saskatchewan was increased during this period, the amount of internal R&D grew at a much faster pace. As a result, the university research program became a progressively smaller fraction of SRC's R&D programs, declining from about 26% in 1957 to about 4% in 1972 (see Table 3.1).

Table 3.1. Illustration of some of the changes experienced by SRC during The Building Years, 1956 – 1972. See also Tables 4.1, 5.1, 6.1.

	1957	**1972**
Total Revenue	$152 k	$1,908 k
Provincial Investment	$124 k	$1,182 k
Contracts	$27 k	$705 k
Fixed Assets (mostly laboratory equipment and fixtures)	None reported	$824 k
Grants (to universities)		
Grant Expense	$40 k	$73 k
Grants as % of Expenses	~ 26%	~ 4%
Employees	15	95
Reference	[37]	[36]

Of course, the building of all these capabilities was a means to an end, that of being able to take on an increasing diversity and depth of programs that would be of strategic interest to the province of Saskatchewan, and to work closely with industry on those programs. In this SRC was in good company in Canada. In a 1971 assessment for the Science Council of Canada Andrew H. Wilson wrote [22]:

"the Research Councils and Foundations in the provinces of Canada are unique. While several other countries have institutes equivalent to our National Research Council, none has quite the same broad range of industrially and regionally oriented institutes in its states, provinces or districts… the six older Research Councils and Foundations appear to have become centres of competence which cannot be ignored. They have growth potential. They are quite effective and flexible in operation, and are generally closer to industry than are the universities. They are in a position to assess regional problems and to form an important link between local industry and the centres of competence in both provincial and federal departments and agencies. They are familiar with the problems of small companies, the more technologically oriented of which have perhaps more innovation and growth potential than most other companies. The Councils are, indeed, a Canadian resource … these institutes have grown and developed to an extent that their competence can be used beyond the borders of their respective provinces and for the benefit of Canada as a whole."

4 THE MATURING YEARS, 1972 - 1983

In May of 1972 SRC promoted Dr. Tom Pepper[27] to be SRC's second Permanent Head [36]. He served in this capacity, as Director of SRC, until early in 1983 [34] (See Appendix 9.4). Dr. Pepper's initial challenges were to evaluate what other kinds of work could be useful to the provincial government and to lead a special effort to increase the amount of work that could be done by SRC on behalf of industry [36]. Another emphasis was placed on identifying and developing processes and products that could enable the development of secondary industries in Saskatchewan [51]. In 1978 The Research Council Act was amended to broaden SRC's mandate to include:

"... *matters pertaining to research, development, design, consultation, innovation, and investigation in the natural and management sciences, pure and applied, as they affect the welfare of the province ...* " [56][28].

One of the first successes in diversifying contract revenue was in winning more contracts with the provincial government, through the Department of Industry and Commerce, and also with the federal government, through the Department of Supply and Services [49]. On the industrial side, the major continuing program areas of focus were [51,52]:

---

[27] Dr. Pepper had been the head of the Physics Division since it was formed in 1958, and had served as the Assistant Director since 1968.

[28] Formerly, it read "... *matters pertaining to research and investigation in the fields of the physical sciences, pure and applied, as they affect the economy of the province ...*"

o Agriculture, with a focus on intensification and diversification of agricultural production capabilities, including feedlots,

o Resource Development, including Northern resource development, minerals, petroleum, and surface- and groundwater,

o Industrial Development, including a wide range of mining, milling, and manufacturing processes, particularly those involving lignite, potash, and clays, plus transportation mechanisms, including pipelining,

o Environmental Protection, including monitoring, analyses and environmental impact assessments[29], and

o Services, including testing and analyses (particularly chemical analyses), and technical information and industrial engineering services, including partnering with NRC on the Technical Information Service (TIS) program (see Chapters 2 and 3), and later the Industrial Research Assistance Program (IRAP), beginning in 1981 [55].

Figure 4.1 SRC's second Permanent Head, Dr. Tom Pepper took office as Director in 1972 [36].

---

[29] Including for such developments as hydroelectric power generation near Nipawin and the coal-fired power generation plant near Poplar River.

Figure 4.2 In 1972 SRC updated its company logo.

As was the case during *"The Building Years,"* projects and revenues continued to expand in this era. In particular, the sources of revenues continued to diversify beyond the core provincial investment. The other sources of revenue continued to grow faster than the core provincial investment level, and became the largest source of revenue for the first time in 1974, comprising provincial government contracts (24%), federal government grants and contracts (20%), and industry contracts (11%) [52]. With contract revenues meeting and then exceeding the provincial investment level it became increasing challenging to achieve appropriate balances between the short-term-focused R&D projects funded by the contracts and the long-range, more strategic R&D projects which were generally funded from the provincial investment.

SRC's growth brought two other critical challenges: management and facilities. The rising levels of work led to the hiring of additional staff, which had risen to 150 by 1978, 174 by 1979, and 200 by 1980 [53,54]. With this level of staffing, Dr. Pepper made it a priority to strengthen the sophistication of management practices at SRC in order to position the company for further growth. As SRC continued to focus on client and industry needs it began to increasingly organize and report on its work in terms of program areas rather than technical disciplines. In 1982 the principal disciplinary organizational structure was replaced with a mostly sector-based structure, including Divisions for Resources, Industrial Technology Transfer, Environment, Services, and the Canadian Centre for Advanced Instrumentation (see below) [55].

The expansions in work and staffing also put a strain on the facilities available in the two Saskatoon buildings, and by 1974, cramped conditions in the main laboratories building had led to the need to expand into a fleet of office and lab trailers (Figure 4.4).

By 1978 a number of additional facilities were acquired by extending the Annex Building (Figure 4.5; behind the main building at 30 Campus Drive), expanding the Engineering Building on Quebec Avenue, and constructing a test facility for residential construction methods on university land at 115th Street [56,57].

In 1979 it was decided to expand again and designs were developed for two new facilities in the new Saskatoon Research Park[30] [53]. Between 1980 and 1981 SRC moved into the Resources Research Facility (RRF), which housed the analytical chemical laboratories, the new SLOWPOKE-2 nuclear reactor, and a multi-purpose pilot plant, and into the SEDCO Centre, which accommodated the Industrial Services Division and parts of the other divisions (Figures 4.6 – 4.8) [58].

Among the expansions was the library, which by 1974 had holdings of 400 periodicals, 7,000 books, and 10,000 other documents, and dealt with 2,000 requests for special documents [52]. By 1979 the library held 10,000 books, and 12,000 reports [53].

With the completion of the moves of staff and facilities into the RRF and SEDCO facilities in 1981 SRC found itself housing staff and facilities in seven distinct locations. This led to planning to increase the amount of consolidation in the Saskatoon Research Park, which was taken up in the next era (see Chapter 5).

---

[30] Now known as Innovation Place Saskatoon.

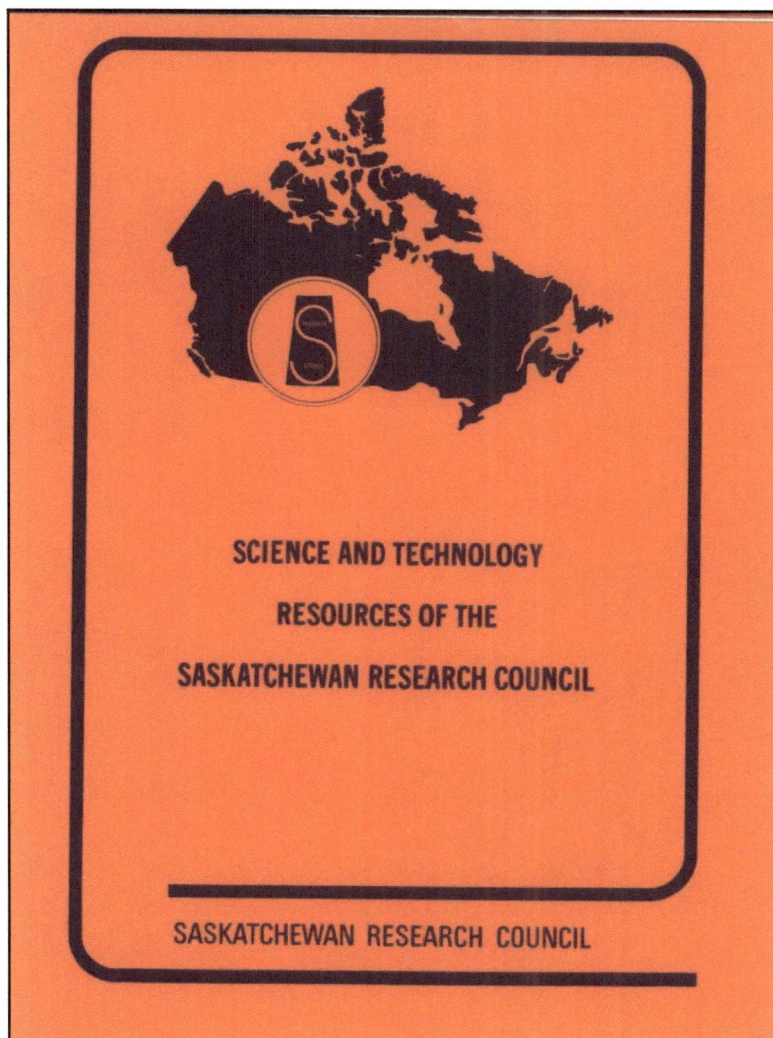

Figure 4.3. SRC published a comprehensive listing of professional staff and expertise areas in 1977.

Figure 4.4. Expansion trailers in 1974 [52].

Figure 4.5. The SRC Storage Building before it was expanded in 1978. Saskatchewan Research Council photo (1963).

Figure 4.6. In 1980 and 1981 SRC moved into new facilities in Innovation Place, Saskatoon. The Resources Research Facility (RRF), on the left, housed the analytical chemical laboratories, the new SLOWPOKE-2 nuclear reactor, and a multi-purpose pilot plant. The SEDCO Centre, on the right, accommodated the Industrial Services Division and parts of the other divisions.

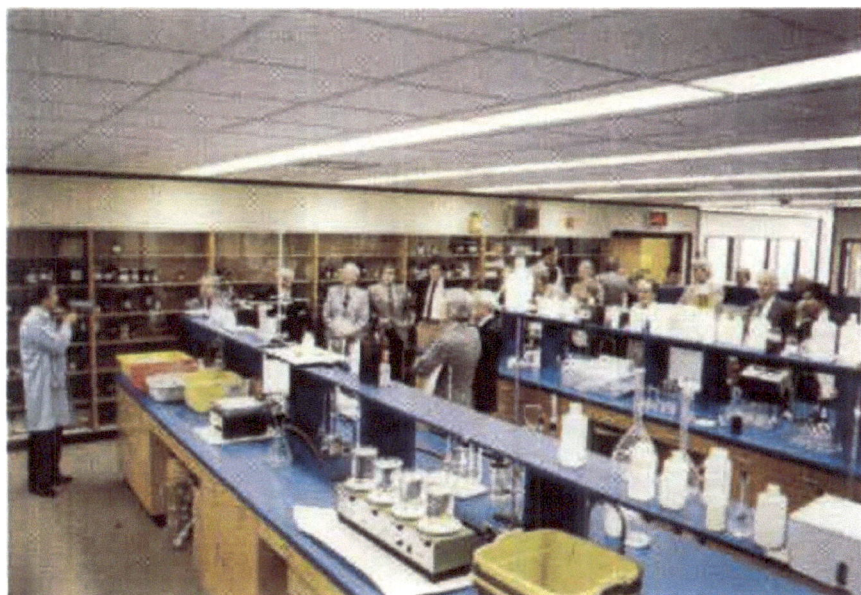

Figure 4.7. Premier Blakeney and other MLAs touring the new analytical chemistry laboratories in 1981 [58].

Figure 4.8. A solvent extraction pilot plant for uranium ore processing, in the new multi-purpose pilot plant facility in 1981 [58].

One of the major new tools acquired in the latter part of *"The Maturing Years"* was the SLOWPOKE-2 nuclear reactor, which was purchased from Atomic Energy of Canada. The reactor was commissioned[31] and went critical on March 13, 1981 (see Figures 4.9 and 4.10). This is a small, pool-type "20kW" reactor of Canadian design that uses enriched uranium fuel, a light water modulator and beryllium reflector to produce a thermal flux of 0.5×10^{11} to 1.0×10^{12} $n.cm^{-2}.s^{-1}$. The reactor is housed in a closed container suspended in a pool of water. Up to 10 pneumatic irradiation transfer systems enable the transport of sample containers into the reactor to be irradiated and then returned for further testing. For more details see reference [59]. The reactor's ability to provide neutron activation analyses provides a means to quickly and non-destructively detect and analyze for a range of environmental contaminants, and its ability to provide delayed neutron counting (DNC) provided a means for rapid, non-destructive analyses for uranium in both solid and liquid samples. Demand for these services was high, and several thousand samples were analyzed in the first six months of operation [58]. The addition of the SLOWPOKE reactor

[31] Under an operating licence from the Atomic Energy Control Board.

further enhanced SRC's analytical chemistry laboratories as one of the best equipped and most versatile in Canada.

In the 1970s, Saskatchewan's modern-era uranium mines and mills started sequentially coming on-stream. This produced rapid increases in industry demand for radiochemical analyses beyond delayed neutron counting. In response, SRC increased its range of radiochemical analytical techniques and rapidly increased its analytical throughput capacity [56]. In addition, a separate Geochemical Analysis Laboratory was established in 1977 to conduct custom analyses of geological samples (rocks, soils, and sediments). Most of the early demand was for uranium analyses, with over 30,000 such samples were analyzed in each of 1977 and 1978 [56,57]. By 1979 this group was supporting gold exploration as well [53].

With the rapid increase in radiochemical and uranium analytical work, it became necessary to add a radiation safety component to the safety program in the late 1970s [53]. This, in turn, led to SRC providing the uranium mining industry with assistance in evaluating their worker's exposure to airborne alpha-ray emissions in uranium mines and mining equipment [58]. In another offshoot of these developments, SRC began working with the uranium industry on decommissioning studies beginning with Eldorado Nuclear's uranium mine and mill in 1983 [55].

Figure 4.9. A model of the SLOWPOKE-2 nuclear reactor showing the placement of the fuel, light water modulator, and beryllium reflector assembly near the bottom of the pool. Saskatchewan Research Council photo, 1981.

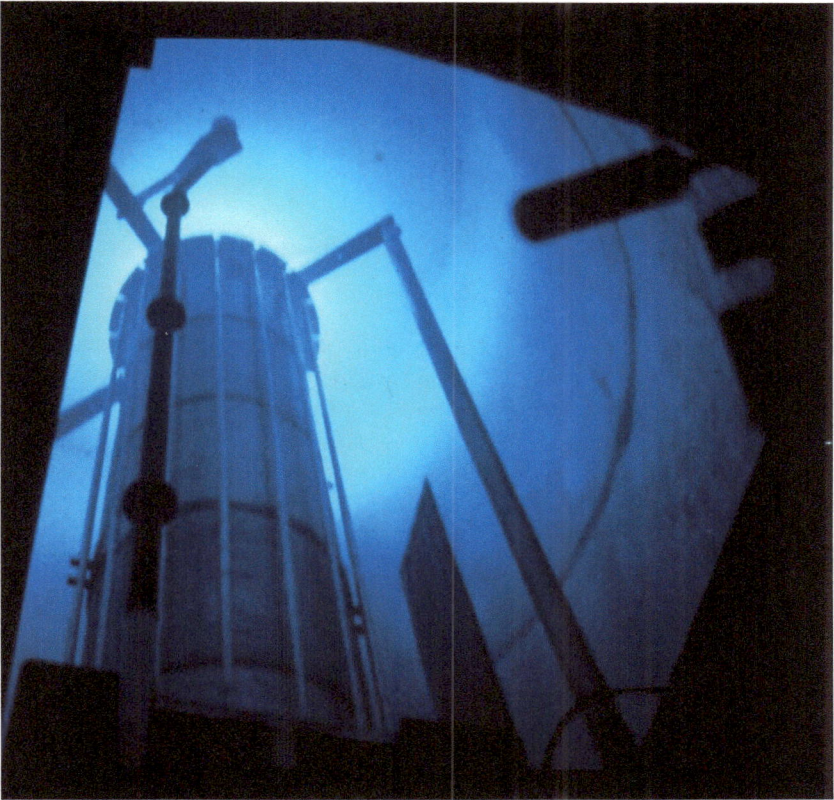

Figure 4.10. Cherenkov glow coming from the base of SRC's SLOWPOKE-2 nuclear reactor. Saskatchewan Research Council photo, 1981.

Some of the SRC "*signature platforms*" that continued to evolve or were initiated include:

Agricultural Studies. Technical investigations and support for many aspects of the agricultural sector were conducted, ranging from alternative crops (including cold and drought resistant crops), to alternative crop management practises (including low tillage and zero tillage practises), to animal feeds and feedlot issues, to uses for waste materials like flax straw, to farm machinery improvements, to value-added products (including biofuels from crops and forestry biomass). A large multi-year program of herbicide and insecticide studies included "drift investigations" of the amounts of these agents escaping the intended spray areas and drifting downwind as droplets and/or vapours. This work involved numerous field trials in a variety of locations, with different spraying equipment, and different herbicides and insecticides (such as 2,4-D), and like so many other SRC programs involved designing and deploying specialized SRC sampling and monitoring equipment (Figure 4.11). These studies resulted in recommendations for manufacturers, farmers, and governments alike [36,49].

Groundwater Studies. Another area that was of prime interest to the agricultural sector was, of course, water. By 1983, over two decades of field studies had allowed SRC to outline the major aquifers of the province, place over 50 observations wells, and to build up a substantial database of groundwater and well-water information, including information on water location, depth, quality, and horizontal and vertical flows (Figure 4.12). This was made available to communities and industries alike in order to assist with such activities as locating new water wells for farms and communities, and monitoring potash tailings ponds. Other programs were concerned with such issues as iron bacteria in groundwater sources and well-water samples (Figure 4.13).

Water Treatment. SRC's desalination program, which was aimed at creating low-cost water suitable for household and agricultural use, continued during this area. SRC's freeze-desalination technology, which took advantage of Saskatchewan's cold winters to naturally freeze high salinity water, advanced to field demonstrations in the communities of Armley, Bruno, Blucher, Saskatoon, and Frontier, Saskatchewan (Figure 4.14) [60]. These were successful, and the large-scale demonstration at Frontier contributed significantly to the village's water supply [60]. Subsequent demonstrations at these communities showed that salt content reductions of 50% could be reliably obtained [57].

Mineral Resource Development. SRC continued to conduct substantial mineral exploration programs, especially for uranium in northern Saskatchewan, for coal and potash in southern Saskatchewan (Figure 4.15 and 4.16), and also for base metals copper, lead, and zinc [57]. Meanwhile, SRC increased the amount of work it was conducting in support of mining methods, including the solution mining of potash, and mineral processing technologies. For example, SRC continued to support the development of uranium mines in the Uranium City area of Saskatchewan, including support for Eldorado Nuclear Ltd's. uranium processing plant [62].

Figure 4.11. Monitoring equipment for pesticide spray-drift studies in 1973 [49].

Pipeflow Technology Development. By the end of 1973 the Engineering Division's pipeline group had established itself as a leading centre for R&D in the hydraulic transportation of solid materials (in the form of slurries) and this provided a platform for increased industrial and government contract work, including work for the Canadian Transport Commission [49] (see Figure 4.17). One of the keys to the success of this initiative was the strong partnership that was forged with the University of Saskatchewan research group led by Professor Cliff A. Shook[32]. This partnership enabled a highly productive blend of discovery research, led by Dr. Shook at the university, with applied R&D led by SRC employees. The pipeflow facility continued to maintain 2, 4, 6, 8, 10 and 12 inch (5 to 30 cm) diameter pipeline test loops, most of which were instrumented with gamma ray density profilers, and a dedicated 2-inch diameter pipeline for volatile and explosive mixtures in crude oils. The kinds of solids evaluated for slurry hydrotransport included coal, potash, limestone, mine tailings, iron ores, crude oils, and coal suspended in crude oil[33].

---

[32] For whom SRC's Shook-Gillies High Pressure, High Temperature Test Facility was co-named in 2015. See Chapter 6.

[33] As a potential transportation method for coal.

By 1977 the pipeline group had begun working on the hydraulic transport of oil sand tailings from Great Canadian Oil Sands and Syncrude Canada Ltd. for the first time [57], beginning programs and relationships that would develop and continue to the present day.

Figure 4.12. Test-hole drilling to determine the areal extent of the Hatfield Valley Aquifer, Saskatchewan, in 1981 [58].

Figure 4.13. Studying the temperature-dependence of iron bacteria growth in ground- and well-water samples, in 1973 [49].

Figure 4.14. Large-scale freeze-desalination demonstration at Frontier, Saskatchewan in 1975 [60].

Ceramics Research Facility. Canada's only Ceramics Research Facility was established at SRC in the early 1970s in order to help industry identify and develop prospective ceramics products, such as bricks (Figure 4.18), tiles, pipes, cements, and granular ceramic products such as cat litter. By 1973 a Saskatchewan company was manufacturing and exporting to the U.S. a white-facing brick developed for them by SRC [49].

Climate Reference Studies. SRC's climatology research continued at the *"Climatological Reference Station"* (Figure 4.19) and more instruments continued to be added. Climatology research was also pursued elsewhere, including through the use of a tethersonde sounding balloon to help develop and validate an atmospheric dispersion model [58]. In 1979 a Lightning Detection System was installed in a northern forest to enable early deployment of fire suppression measures [53].

SRC's climate studies eventually led to opportunities to do more than just study the climate and its variations. In 1974 and 1975 SRC's climate-related work was expanded into evaluations of wind-power potential for the Saskatchewan Power Corp. (now SaskPower) [52,60]. See also Chapters 5 and 6.

Remote Sensing and Imaging. SRC began collecting aerial photos in the 1960s, including historical aerial photos dating back to 1928, had an extensive collection by 1972, and produced photomosaics compatible with the National Topographical System and covering the entire province (Figure 4.20) [36,49]. In 1974 SRC's work in remote sensing and imaging was extended to the use of both aircraft and satellites[34] plus the necessary surface observations ("ground-truthing") needed to establish interpretation methods [52]. This was extended into the mapping of forests (Figure 4.21) and forest fires using data from meteorological sensors on the NOAA-2[35] and subsequent NASA satellites [60]. By 1981 SRC had the only known complete collection of 1:50,000 scale photo mosaics covering the entire province [58]. This program expanded in *The Commercial Years* (Chapter 5), and morphed into the Geospatial Imagery Collaborative in *The New Millennium Years* (Chapter 6).

---

[34] In cooperation with the Canada Centre for Remote Sensing.
[35] The National Oceanic and Atmospheric Administration satellite NOAA-2 was mainly used for cloud imagery and temperature profiling; see also Appendix 9.11.

Figure 4.15. Drilling to evaluate underground coal resources near Boundary Dam in 1974 [52].

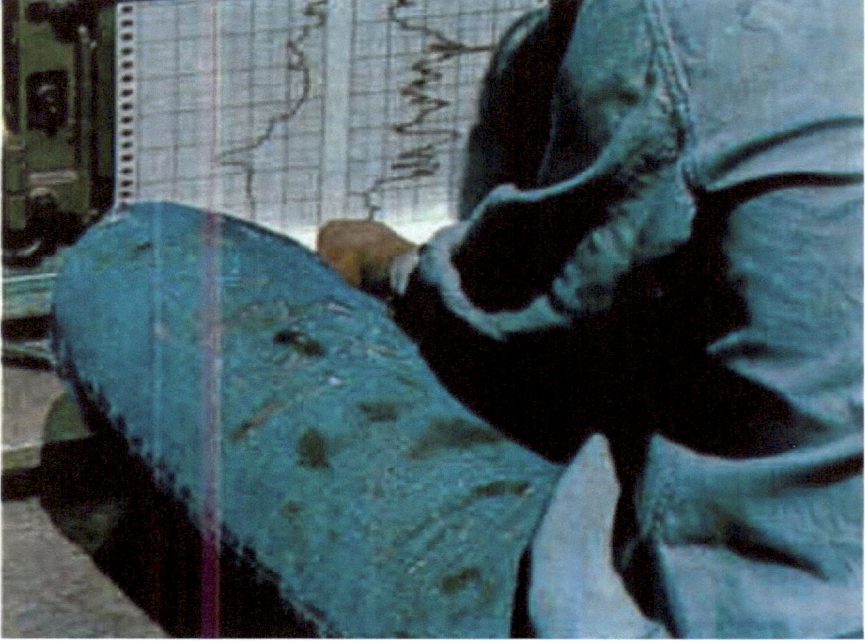

Figure 4.16. Electric logging while drilling to evaluate underground coal resources in 1974 [52].

Archaeology. Although SRC had been engaged in carbon-14 dating of materials since the 1950s, it had been mostly for geological studies. In the 1970s, SRC's carbon-14 dating capabilities were increasingly used to support archeological studies, such as those of the National Museum of Man [49]. In 1976 a formal archaeology group was formed in order to support environmental impact studies for potential construction projects such as proposed dams and hydroelectric and thermoelectric installations (Figure 4.22) [57,61,62]. Over the years, archeological studies were conducted across the full extent of the province, South to North and West to East [61]. Such projects led to the discovery of a number of historical sites ranging in age from 8,000 years ago to the 1700s [62]. For example, significant archeological sites discovered and examined in conjunction with the Nipawin Reservoir hydroelectric project included distinct sites dating back 5,000 years and 3,000-4,000 years, plus a site from the 1400-1700s, and fur-trading posts from *ca.* 1770 and *ca.* 1850 [63]. Such work continued through the 1980s[36].

---

[36] In 1990 the archaeology team was spun-out and privatized as Western Heritage Services Inc. (see Appendix 9.7).

Figure 4.17. SRC's pipeflow centre in 1973 [49].

Stack Sampling and Monitoring Studies. In 1973 particulate emission studies were conducted for a feed mill, a potash mine, and a grain elevator [49]. Over the years this work continued and expanded. In 1975 SRC started to monitor and model the dispersion of dust plumes into the environment [60]. In 1978, at the request of the potash industry, SRC created a team to monitor particulate emissions from large industrial stacks and to help the industry with regulatory compliance (Figure 4.23). In 1979 this program was expanded to include radon, emissions from coal-fired power plants, a broader range of pollutants from a range of industries, and dust issues in barns, grain elevators, farm equipment, and potash mines [53].

Figure 4.18. Colour matching prototype clay bricks for an industrial client in the SRC Ceramics Research Facility in 1972 [36].

Some of the new program areas that emerged in the 1970s included:

Winter Ice Road Studies. Numerous Saskatchewan communities rely on winter ice roads[37] for transportation needs that are met by ferries during ice-free parts of the year, but ice road safety is a huge concern. In the 1970s SRC developed a *"deflectometer"* (Figure 4.24) as part of an investigation, for the Department of Highways and Transportation, into the pressures produced by vehicles moving at different speeds across ice roads having different ice thicknesses [60,62]. This work was used to help determine safe vehicle mass and speed limits for winter ice roads having different ice thicknesses.

---

[37] In the 1970s these were sometimes referred-to as ice bridges.

Figure 4.19. Maintaining a Sunshine Recorder at the SRC *Climatological Reference Station*" in 1980 [54].

Figure 4.20. SRC had an extensive collection of aerial photographs by 1972 [36].

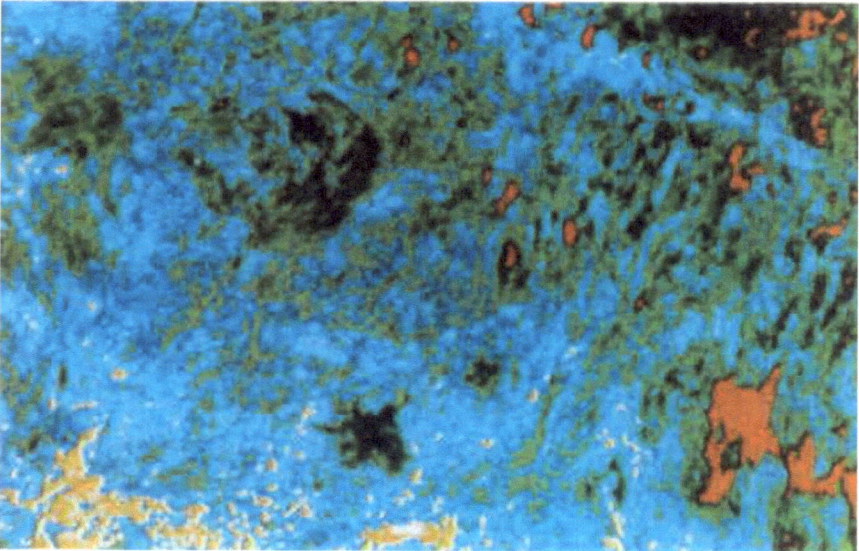

Figure 4.21. An early example of a satellite image with overlaid colour-coding to differentiate wood species in the Clearwater River Valley forest (1978, [56]).

Figure 4.22. An archeological dig conducted as part of an environmental impact study for a potential industrial project in 1977 [57].

Figure 4.23. Monitoring stack emissions from a potash plant in 1978 [56].

Energy Conservation Programs. By 1976 SRC had started working in the area of farm and building energy efficiency and energy conservation. One such program focused on farm energy conservation through such means as saline-resistant crops, increased crop productivity, biofuels development, and even waste biomass energy production. Another program focused on energy conservation in buildings, targeting both residential and farm homes and buildings [57,62]. An early success was the 1977 *"Saskatchewan Conservation House"* (Figure 4.25), which demonstrated efficient methods for energy conservation and heating, including solar heating. With this demonstration house, SRC was able to show that practical insulation and conservation practises could reduce home energy requirements by up to 90% [56]. SRC was later co-recipient of a Conservation Housing Design Council Award of Merit, in 1982 [63]. In a related program area, SRC established six retrofit demonstration houses in Yorkton, Moose Jaw, Swift Current, Weyburn, Prince Albert, and Melville. The retrofit houses demonstrated that 40% energy savings could be easily achieved [54]. The learnings from these and other conservation projects were freely disseminated to the home building industry and others[38].

---

[38] SRC was also an active supporter of Saskatchewan's Energy Conservation Information Centre.

Figure 4.24. Measuring the ice thickness (upper) and pressure generated by vehicles (lower) on winter ice roads across Lake Diefenbaker, near Riverhurst, Saskatchewan in 1975 and 1976 [60].

Figure 4.25. SRC's "*Saskatchewan Conservation House*" in 1977. This house demonstrated efficient methods for energy conservation and heating, including solar heating.

Plastics and Manufacturing. In 1972 SRC developed an injection moulding press (Figure 4.26) for manufacturing plastic components for instruments of all kinds, but initially for environmental air and water monitors [36,49]. Although most of these instruments were used for SRC programs some were sold to the federal Department of the Environment as well. This capability was also used to investigate the economics of short-run plastic parts production for the manufacturing sector [49].

Highway Test Track. In 1975 SRC designed a 12m diameter, circular indoor test track for the accelerated testing of new highway materials and construction designs for the Department of Highways and Transportation. This was the first of its kind in Canada [60]. Construction was initiated in Regina in 1976 (Figure 4.27) and completed in 1978, and SRC continued to provide technical support for the facility [53]. In 1995 it was transferred to the Canadian Transportation Innovation Centre (C-TIC) with which SRC continued to collaborate [64] (see also Appendix 9.7).

Figure 4.26. SRC injection moulding press for manufacturing plastic components. Saskatchewan Research Council, 1973.

Figure 4.27. The Highway Test Track Facility taking shape in 1976 [62].

In many ways, these were *The Maturing Years*. SRC not only continued to build dedicated staff, equipment, and facilities to closer to *"critical mass,"* but the company really learned how to listen to industry and community needs, conduct focused applied R&D projects, and deliver practical results within reasonable time periods. SRC increasingly pursued its own fee-for-service revenue and client-needs became increasingly important. With increased market focus SRC's work began to diversify, and increased attention was given to small- and medium-sized enterprises (SMEs). In December 1976 SRC established a subsidiary, Sareco Holdings Ltd.[39], as a vehicle to create and own satellite companies for specific technical commercialization undertakings [57]. By the end of this era, a dominating focus on clients and contracts had made SRC mostly market-driven in its outlook.

In a 1983 review of the provincial research organizations of the time Le Roy and Dufour concluded that [15]:

"... their common purpose is to make available the fruits of research likely to be beneficial to the provinces and to the nation as a whole, with particular emphasis on industrial development. This they do by undertaking research and development work and by making available technical information, advice, and know-how, as well as by providing analytical and testing services. In assuming this role, they serve as veritable research arms for thousands of small and medium-sized enterprises, an important segment of industry for employment and the creation of wealth in the country. They also serve as significant instruments in realizing both federal and provincial strategies for technological development by maintaining and enhancing technology transfer, industrial productivity and innovation."

These features applied as well to the SRC of the time as to any of its sister provincial research organizations (PROs) in Canada.

The continuing growth of SRC's physical facilities and equipment is reflected in the doubling of fixed assets, doubling of the staff level, and the growing amount of R&D activity, as well as the growth of the total annual revenues, which increased five-fold during this period. The amount of provincial investment tripled (see Table 4.1) while the amount of contract revenue increased by an order of magnitude, to become nearly twice the provincial investment level.

Although the amount of money allocated to sponsoring applied research at the University of Saskatchewan was again increased during this period, the amount of internal R&D grew at a much faster pace. As a result, the university research program once again became a progressively smaller fraction of SRC's R&D programs, declining from about 4% in 1972 to only 1% in 1983 (see Table 4.1).

---

[39] Sareco does not appear to have ever been active, and it became legally inactive and struck from the business register in 1991.

Table 4.1. Illustration of some of the changes experienced by SRC during *The Maturing Years***, 1972 – 1983. See also Tables 3.1, 5.1, 6.1.**

	1972	1983
Total Revenue	$1,908 k	$10,102 k
Provincial Investment	$1,182 k	$3,519 k
Contracts	$705 k	$6,282 k
Fixed Assets (mostly laboratory equipment and fixtures)	$824 k	$1,673 k
Grants (to universities)		
Grant Expense	$73 k	$118 k
Grants as % of Expenses	~ 4%	~ 1%
Employees	95	212
Reference	[36]	[34]

LAURIER L. SCHRAMM

5 THE COMMERCIAL YEARS, 1983 - 2000

Early in 1983 SRC hired Mr. Jim Hutch as SRC's third Permanent Head [55]. He served in this capacity, as President of SRC, until 1995 [34] (See Appendix 9.4). The progressively increasing importance placed on conducting contract work for industry was driven further under a new five-year strategic plan, the *"Long Range Plan,"* which was launched in 1983 and was aimed at focusing SRC's work on the needs of the commercial marketplace [34]. In this new strategic plan, the emphasis was shifted *"from SRC as producer to the client as consumer of technology"* [51,63]. In 1984, the Board of Directors became *"industry led in order for SRC to be more responsive to the marketplace"* [65], and *"bring it closer to industry"* [66], which further contributed to the new strategic direction. Revenue from contracts, particularly industry contracts, now continued to increasingly exceed revenue from provincial investment throughout this era.

Another sign of the increasingly commercial focus was the elimination of the university grants program which SRC had initiated in 1948. This program had been wound down by 1989/90 [76]. Similarly, SRC's in-kind support of the Provincial Government scholarship program (which SRC had initiated in 1949) was suspended in 1988 [67,68].

As mentioned in Chapter 4, SRC's facility expansions of *The Maturing Years* had resulted in the company housing staff and facilities in seven distinct locations, which led to a desire to increase the amount of consolidation in the Saskatoon Research Park (later renamed Innovation Place, Saskatoon, see Figure 5.3). Both a degree of consolidation and some expansion of facilities space were largely accomplished between 1985 and 1986 [69,70].

Figure 5.1 SRC's third Permanent Head, Mr. Jim Hutch took office as President in 1983 [34].

Under the *"Long Range Plan,"* SRC's organizational structure was adjusted somewhat. 1985/86 marked the formation of dedicated Human Resources and Communications functions and the creation of a dedicated Administration and Finance Group [70]. By 1987 the structure had evolved into essentially four main groupings: a Research and Development Branch (with distinct Resources, Petroleum and Environment Divisions), a Technology Transfer and Business Development Branch (containing CCAI), a Scientific and Analytical Services Branch (with Analytical Chemistry and Bovine Blood Typing/Genetics Divisions), and an Administration and Financial Services Branch[40] [70,75].

---

[40] The exact structure of Branches and Divisions changed almost every year during this era.

Figure 5.2. In 1983 SRC updated its company logo twice, first to the image on the upper right, then to that on the lower left [34,71].

Several new programs made relatively brief appearances before being merged into broader groups at SRC, including the Canadian Centre for Advanced Instrumentation (CCAI) and a CAD/CAM Robotics Centre:

Canadian Centre for Advanced Instrumentation (CCAI). In late 1982, with federal funding support, this centre was launched with the aim of providing advanced instrumentation design, development, and services to the natural resource and high-technology industries [63]. In 1983 CCAI developed a prototype automated air sampling system, the MINITUBE™ Air Sampling System (MASS), for the federal Department of National Defense [55] (see Figure 5.4). The MINITUBE™ included a carousel containing 50 minitubes for sample collection and adsorption [55,71]. The MINITUBE instrument earned SRC one of its first three patents [72] (see also Appendix 9.9). Twelve production units were sold to National Defense in 1984 and an additional 38 in 1985 [69]. A few units were later sold internationally, including to the West German military in 1986 [75]. In 1988 diminishing external funding led to CCAI being mostly wound down, with some of its activities being maintained, but re-integrated within other existing SRC Divisions.

CAD/CAM Robotics Centre. In 1983/84, again with federal funding support, the Saskatchewan CAD/CAM (Computer-Aided Design, Computer-Aided Manufacturing) Robotics Centre at SRC was launched [71]. This centre was positioned to assist manufacturing clients with advanced product design and prototyping support. As part of this program, SRC placed CAD/CAM design stations (Figure 5.5) in manufacturing plants throughout the province [71]. This program was operated for several years before being merged into the Product and Process Development Division in 1988.

Figure 5.3. The Galleria Building in Innovation Place, Saskatoon, with SRC as an anchor tenant. Saskatchewan Research Council photo.

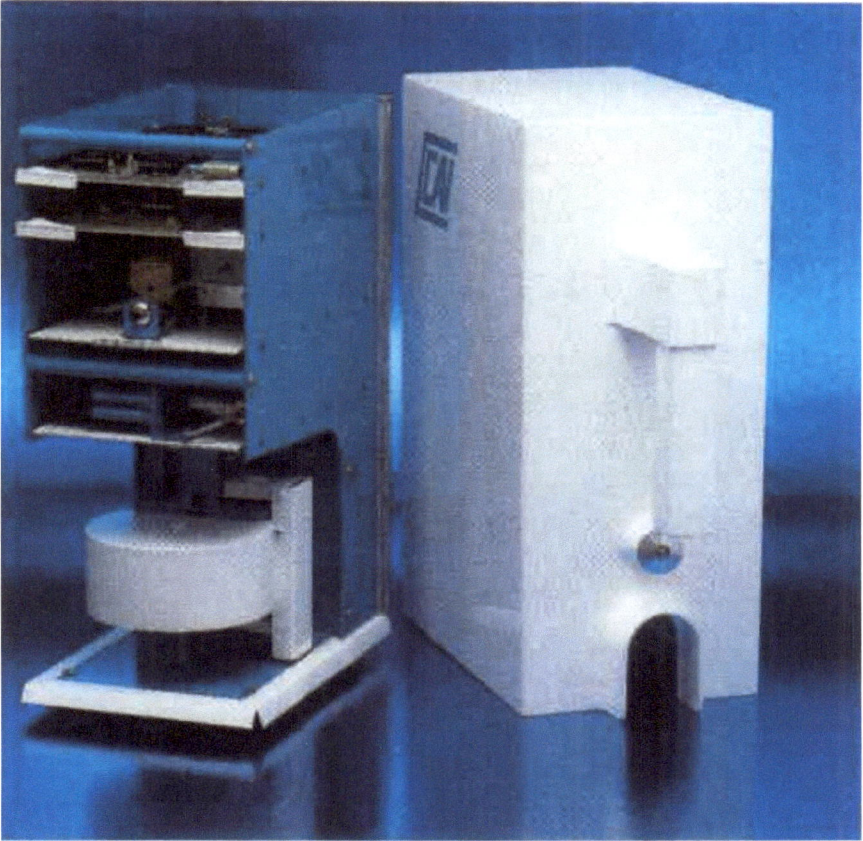

Figure 5.4. The MINITUBE™ Air Sampling System (MASS) in 1984 [71].

In 1987 the Prairie Agricultural Machinery Institute (PAMI), a Humboldt, Saskatchewan-based research and technology organization (RTO) focused on applied research, development, testing, and evaluation related to farm machinery, was on the verge of discontinuance due to federal and provincial funding reductions. SRC was asked to assume responsibility for the management of PAMI, and developed and implemented a five-year strategic plan to reposition the organization for increased fee-for-service contract work [67,78]. SRC ultimately managed PAMI under contract from 1988 through 1995, at which time it was spun-out as an autonomous organization (see also Appendix 9.7).

Figure 5.5. A CAD/CAM computer in 1986 [75].

One of the principal features of *The Commercial Years* is the substantial business acquisitions that were made in the areas of petroleum, bovine parentage, and building performance. These acquisitions strengthened and diversified SRC's business offerings and were responsible for most of the doubling in annual revenues that occurred during this period.

Petroleum. On June 1, 1985, SRC acquired the Saskatchewan Oil and Gas Corporation's (SaskOil's) R&D Centre in Regina [69] (Figure 5.6; see also Appendix 9.6). It was decided to keep the Centre in Regina and make it the foundation for a dedicated Petroleum Branch with an initial staff of 31 employees. This marked a return of permanent SRC facilities in Regina for the first time since 1958. The SaskOil R&D Centre acquisition strengthened SRC's expertise, facilities, and capabilities in the sub-sectors of conventional light oil, conventional heavy oil, and natural gas, especially in the enhanced heavy oil recovery process area (Figure 5.7). The acquisition was also significant in that it brought SRC into significant active engagement with every major resource in Saskatchewan.

The new Petroleum Branch moved quickly to develop a rapid-response Mobile Emulsion Treating unit, designed to be moved to producing oilfield locations on demand to help find solutions to oilfield production problems such as the breaking of wellhead emulsions (Figure 5.8). By 1995 a number of what would become long-standing enhanced oil recovery (EOR) R&D program areas had been established in gas and chemical flooding, steam flooding, *in situ* combustion, horizontal well technologies, and field upgrading [73,74]. One of the gas flooding programs that was launched [64], involved working with PanCanadian Petroleum Ltd. to develop a carbon dioxide (CO_2) flooding EOR process for the Weyburn reservoir in southern Saskatchewan that would become one of SRC's largest economic impact achievements (see Chapter 7).

Figure 5.6. The new Petroleum Branch building in Regina, in 1985 [69].

Figure 5.7. Part of a core displacement apparatus used to study enhanced heavy oil recovery processes (1985) [69].

Figure 5.8. SRC engineer Jerry Scoular operating SRC's Mobile Emulsion Treating unit in 1986 [70].

Bova-Can and GenServe™. In 1987 SRC acquired the Bovine Blood Typing Laboratory from the federal government[41] and moved it from Ottawa to Saskatoon [75]. The Lab's first leader, Dr. Gerry Kraay also relocated from Ottawa to join SRC. The initial team comprised seven people and their initial focus was to provide *"cattle producers across Canada with a comprehensive testing service for maintaining breed purity"* [75]. The service was offered to both breeders and their industry associations. Initially, the service was based on parentage testing by blood typing (Figure 5.9), with nearly 20,000 animals being blood typed in 1986/87 alone [75]. Almost immediately upon establishing the lab in Saskatoon, the group launched a research project on the possibility of using DNA fingerprinting for unambiguous parentage testing [75]. This group's work was conducted with close connections to the cattle breeding industry, through Bova-Can Parentage Testing Inc., an advisory board made up of several national breed associations.

A number of additional services were developed over the next few years. Commercial parentage testing based on DNA was offered to the industry beginning in 1988 [68], Karyotyping was offered in 1989 (which can detect certain chromosomal abnormalities in cattle, [76]), as were parentage testing services for elk (which were later extended to all species of domestic animals except horses [77]). Research aimed at a possible market for DNA testing for other animal traits was also initiated in 1989 [76], which led to additional DNA-based trait tests being offered in subsequent years [79] and the launch of a national DNA sample bank in 1992 [77].

This group evolved into Bova-Can Laboratories, a joint-venture partnership between SRC and Bova-Can Parentage Testing Inc. in 1996 [74]. Bova-Can's expanded mandate included providing applied R&D and high-quality genetic testing and interpretation services to Canada's livestock industries. In 1998 a second but related group, GenServ Laboratories, was launched to provide DNA-based services involving plants [78]. These services included testing for the presence and identity of genetically modified organisms (GMOs).

---

[41] In 1986 the federal Nielsen Commission had recommended that the blood typing lab in Ottawa be separated from the federal government ("An introduction to the process of program review," Supply and Services Canada, Ottawa, 1986).

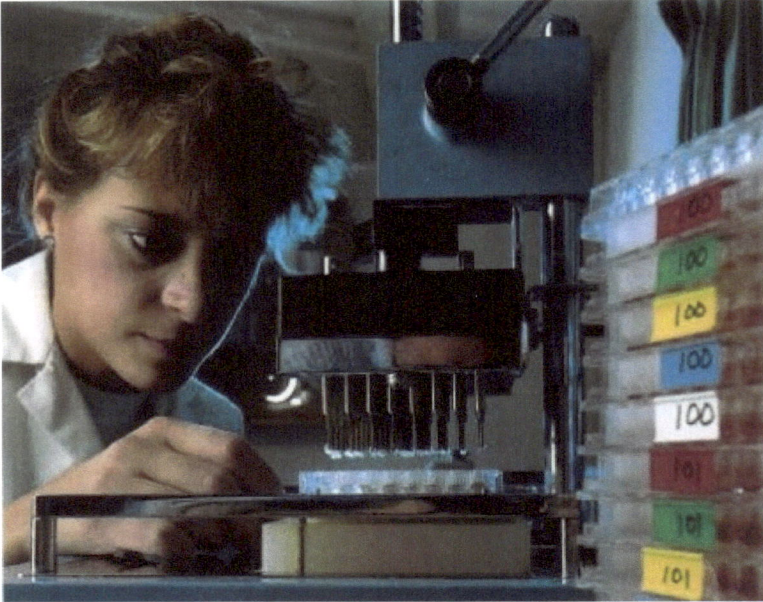

Figure 5.9. Blood-based bovine parentage testing in 1987 [75].

Building Performance and Energy Conservation Programs. Based on SRC's energy conservation program research two important practical guides were published in order to disseminate the results to Canada's home-building industry: "Energy-Efficient House Construction" was published by Canada Mortgage and Housing Corp. (CMHC) in 1984, and "Air Sealing Homes for Energy Conservation" was published by the federal Department of Energy, Mines, and Resources in 1984. Both of these publications received awards for excellence in technology transfer [69]. In subsequent years a steady demand built-up for building energy audit services [78].

In 1990, SRC acquired the Building Science Division from the National Research Council and transferred it from Ottawa to Saskatoon [51] (see also Appendix 9.6). This addition enabled SRC to expand its already successful energy conservation program related to housing expanded to other kinds of buildings, including schools and recreational facilities [71]. The main programs in this area included building energy monitoring, energy efficiency, building envelopes, air tightness testing, and computer modelling of building energy systems. This work complemented and expanded on what SRC had already been doing in indoor air quality testing (Figure 5.10), including radon testing [79,80].

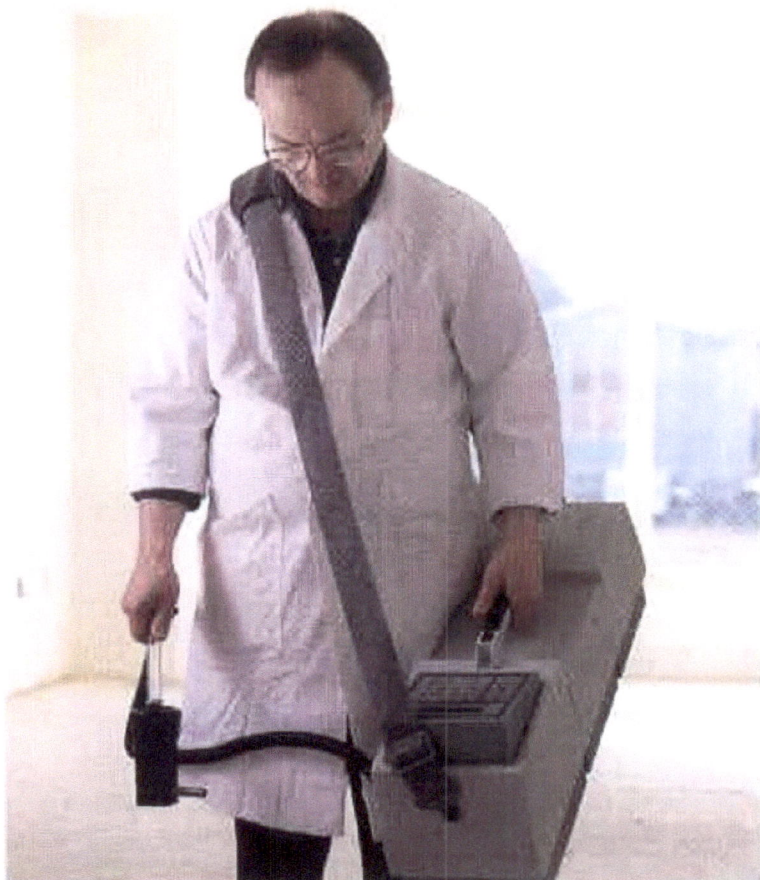

Figure 5.10. Using a multigas infrared analyzer to test indoor air quality in 1991 [79].

In 1991 SRC commissioned a large-scale (23 m³) environmental chamber for the evaluation of such material properties as coating performance, cleaning techniques, and formaldehyde off-gassing from composite wood products [77,80]. This was the first such test chamber of its kind in Canada. In subsequent years this group provided national leadership in such areas as the federal Advanced Houses Program and the National Energy Codes for Houses [65]. SRC was one of the contributing partners in the Saskatchewan Advanced House demonstration in Saskatoon (Figure 5.11), which achieved a 50% reduction in water consumption, 75%

reduction in space heating, 75% reduction in space cooling, and a 50% reduction in other electricity use [77,81] (see also Section 7.5 below). Later in the 1990s SRC also developed and applied a uniquely prairie expertise in energy efficiency and energy conservation practises for community ice rinks [74].

Figure 5.11. SRC was one of the contributing partners in the Saskatchewan Advanced House demonstration in Saskatoon. At its grand opening in Saskatoon, in 1993, it was billed as *"One of the Most Energy-Efficient, Environmentally-Friendly Homes in the World"* [81]. Photo courtesy of Derek Verhelst, Kelln Solar.

In 1999 SRC employee Dr. Rob Dumont won a national award for designing, building, and demonstrating a leading-edge energy efficient house[42] [82]. In this case, Dr. Dumont used his family's personal residence as the test/demonstration site of the same technologies with which SRC was helping local industry and communities.

---

[42] See also SRC's national award for the "Saskatchewan Conservation House" in 1977, in Chapter 4; and see also Section 7.5 below.

Some of the SRC *"signature platforms"* that continued to evolve or were initiated include:

Agricultural Studies*.* Among the main agricultural programs was one on farm energy management, under which significant energy conserving practices were identified and field-demonstrated on 22 Saskatchewan farms [69]. It was found that with the appropriate combinations of crop, reduced tillage, and snow management, a return of about 50% more grain per unit of fuel and fertilizer energy could be achieved [69]. Another substantial program focused on alternative crops, such as the potential for Kochia as a high-yield forage crop with drought resistance and salinity tolerance [68], and such as agroforestry systems using multiple-row shelterbelts in northern Saskatchewan [79].

SRC's ground water programs continued, including maintaining the network of 50 observation wells (each providing coverage of a 5,000 km^2 area) that had been established previously across southern Saskatchewan [71].

By about 1988 a serious effort had been developed to evaluate value-added products from crops and crop by-products (Figure 5.13). This was a broad program that considered feedstocks such as crop residues, forages, and oil seeds, and potential products including foods, fuels, fibres, and feedstocks [68]. The fuels from vegetable oil technology developed by SRC was patented in the 1990s [83-86] (see also Appendix 9.9).

"Would you believe I was commissioned to do a cost study on pipelining wheat? Using liquid nitrogen as a solvent! I did this study – I remember the insulation on the pipe was very thick – but the total cost of the project was $15 billion! When I discussed that this was the figure, it was just kind of...dropped." (On a proposal to use slurry pipeline technology to transport wheat to the west coast) – *Retired employee*

Figure 5.12. Not all research projects turned out to be practical. The above reflection is from a retired SRC employee in 1997 [74].

Figure 5.13. Analyzing chaff samples for fermentable sugars in 1988 [68].

Mineral Resource Development. SRC's work in the mineral exploration area included systematically studying the geochemistry, mineralogy, and geology of Saskatchewan's metallic minerals, in order to develop models that can be used aid in mineral exploration and development. The principal continuing mineral resource development programs focused on gold and uranium exploration, and also programs dealing with uranium mine tailings, including the possible recovery of gold and radium from these tailings [69,71].

In 1994/95 SRC launched a uranium exploration program termed Wollaston EAGLE, with the support of four uranium companies that were active in the province at the time [73]. The purpose of the program was to develop new exploration tools and techniques to assist the uranium exploration industry, with an initial focus on the Wollaston Lake region of Saskatchewan. This program led to the development of a geological model of the environment needed for uranium deposition and enrichment, which in turn was widely used to help the industry plan focused and cost-effective uranium exploration programs [78,87].

Geochemical services continued to be in demand, mostly from the uranium industry (~80%) and secondarily from the gold industry (nearly 20%) [70]. In 1988 and 1989 SRC began providing services in support of diamond exploration, and other precious metals like platinum and palladium [68,76]. Over 50,000 rock and mineral samples were analyzed in 1989/90. An early highlight was finding one of the first macro diamonds in Saskatchewan for Uranerz Exploration (which they named "*snowball*," as it was as "*white as snow*"). Similarly, process development and demonstration at the laboratory and pilot plant scales continued to be in demand, especially for uranium and gold producers. By 1990 SRC had a diamond processing pilot plant in operation as well [79]. SRC's bulk caustic fusion method for diamond analyses, which involves using sodium hydroxide to dissolve away virtually all of the rock matrix saving the diamonds, was developed in the 1990s and became a world-standard technique [73].

Pipeflow Technology Development. Continuing pipeline programs were related to coal transportation, gold and potash mine tailings, oil sand hydrotransport, oil sand mine tailings, and heavy oil transportation options (including oil-in-water emulsions and oil-diluent mixtures) [69,71,73], see Figure 5.15. By 1999 yet another expansion was needed to keep up with industry needs so a new, and much larger, facility was designed and constructed on 51st St. in Saskatoon [82] (Figure 5.16).

Figure 5.14. Separating heavy mineral fractions from till samples in support of industrial gold and diamond exploration projects (1990) [79].

Figure 5.15. Operating SRC's 20" (50cm) diameter pipeline loop in 1984 [69].

Figure 5.16. The new and expanded Pipe Flow Technology Centre in 2000. Saskatchewan Research Council photo.

Environment. In the 1980s SRC began to bundle its distinctly environmental groups and programs under a single broad "Environment" grouping, including programs in air, water, and soil (Figure 5.17 – 5.22). Programs in aquatic biology and terrestrial ecology continued to grow, while programs in atmospheric processes (including the Climate Reference Centre) were maintained[43].

As its aquaculture capabilities and facilities expanded, SRC was able to undertake hatchery programs related to rainbow trout, arctic char, brine shrimp, and crayfish [69,70]. An aquatic toxicology laboratory was launched in 1986 [75] and expanded over the years to enable a broad range of services for both industry and government.

Similarly, the programs in water quality and supply, air emissions and air quality, climate reference monitoring, waste and land management continued to grow. In the area of water quality research, SRC established a long-term water quality collaboration with the Prairie Farm Rehabilitation Administration (PFRA) that would endure for nearly two decades.

---

[43] In 1998 Dr. Elaine Wheaton published a book entitled: *But It's a Dry Cold: Weathering the Canadian Prairies*, which describes the weather conditions typical of the prairies: blizzards, dust storms, tornadoes, hail, and extremes of heat, cold, and drought. See Appendix 9.8.

Figure 5.17. Soil sampling to evaluate the effects of oil and gas operations on cropland in 1987 [67].

In 1989 SRC increased its level of climate change research [76], fuelling a trend that would see this become another "*signature program*" for SRC, with substantial attention being paid to both the evolution of climate change, the impacts of climate change, and also on ways to help industries, communities, and governments minimize the negative impacts and maximize the positive impacts of climate change [79]. By 1999 SRC was collaborating in this area with numerous partners under the Prairie Adaptation Research Cooperative (PARC) [82].

Numerous remote sensing projects were undertaken with the federal Department of Energy, Mines, and Resources and the Canada Centre for Remote Sensing (CCRS). These typically involved demonstrating and transferring technologies related to the collection, management, and uses of remotely-sensed satellite data (geographic information systems, GIS). The GIS application areas covered included forestry, water, wildlife habitats, mineral deposits, agriculture, and wetlands [68,70]. By 1993 SRC had launched the South Digital Land Cover Project and the North Digital Land Cover Project, with many collaborating organizations across Saskatchewan [65].

Figure 5.18. Installing a piezometer for the study of underground water flows and chemistry (1988 [68]).

Figure 5.19. Winter water sampling in 1988 [68].

By 1995 a wide range of satellites was being accessed, including LANDSAT, NOAA, SPOT, and ERS-1[44], and custom GIS projects were being undertaken for individual industry clients [73], see Figure 5.23.

---

[44] The National Aeronautics and Space Administration's LANDSAT satellites have provided the longest-duration acquisition of satellite imagery of Earth in history. SPOT is a CNES satellite program providing high-resolution optical-imaging Earth observation. The National Oceanic and Atmospheric Administration satellite NOAA-2 was mainly used for cloud imagery and temperature profiling. ERS-1 was an ESA European remote sensing satellite, and the first Earth-observing satellite program to use a polar orbit. See also Appendix 9.11.

Figure 5.20. An SRC-designed and built dew-point-instrument calibrator used in field studies of evapotranspiration of water from soils (1989 [76]).

Figure 5.21. One of the most recognizable instruments at the SRC Climate Reference Centre, a Campbell-Stokes bright sunshine recorder (1990) [79].

Figure 5.22. Toxicologist Mary Moody conducting toxicity testing in 1998 [78].

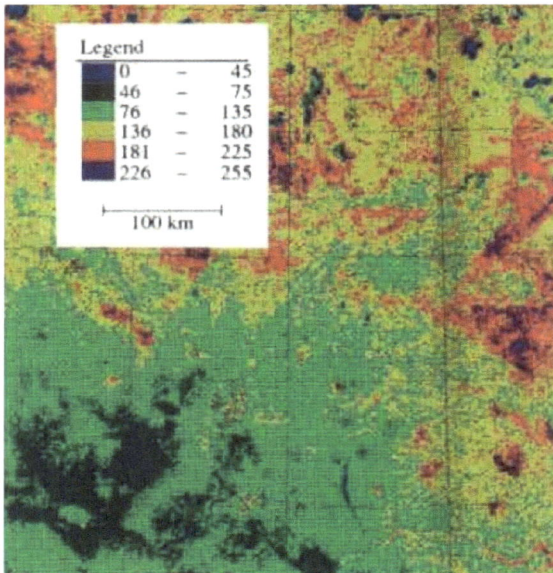

Figure 5.23. This GIS map of southern Saskatchewan was derived from 1988 NOAA satellite data, plus pasture and survey data, to show the relative vigour of vegetation at a time when drought conditions were affecting pasture lands [79].

Some other program areas that emerged or expanded in the 1980s included:

Plastics and Manufacturing. SRC continued to support manufacturers with designs, moulds, prototypes and small-scale production. A highlight of this era was SRC's design of several products for local manufacturers, such as the E-ZEE Wrap 1000 plastic wrap dispenser, including manufacturing the production tooling, both for Jim Scharf Holdings Ltd. in Perdue, Saskatchewan. This and other SRC-designed products product won "Best Product" awards from national bodies such as the Canadian Manufacturing Association, for example, in 1991 [80] and 1992 [77].

Continuing programs related to ceramic clays included work on purifying clays for use in stoneware, and on bleaching clays for potential use as a replacement or imported clays in paper and paint production [71]. Significant efforts in this area seem to have diminished by about 1988.

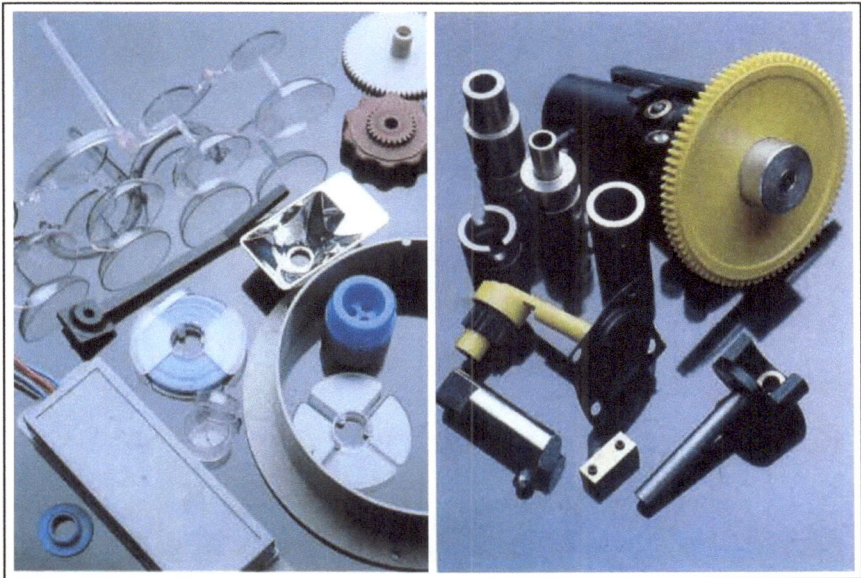

Figure 4.24. SRC continued to support manufacturers with designs, moulds, prototypes and small-scale production (1989, [76]).

Alternative Fuel Technologies. In the 1980s, SRC began developing and demonstrating compressed natural gas (CNG) technologies as a way to provide cost efficient, clean-burning fuel for farms. This included work on trucks, tractors, and on-farm CNG fuelling stations [68]. SRC also developed a natural gas fuelled tractor, and also two dual-fuel tractors[45] (Figure 5.25) that could operate on either conventional diesel fuel or a 50:50 blend of diesel fuel and CNG [68].

In the early 1990s this work had extended to other vehicles, such as the conversion by SRC of Chevy Sprint/Geo Metro vehicles to CNG (Figure 5.26), and assisting SaskEnergy with the conversion of their fleet of about 60 natural gas vehicles [65,80]. By 1998 SRC was recognized as a national leader in the development of natural gas vehicle control systems [87].

Figure 5.25. Photo of an SRC Dual-Fuel™ tractor converted to run on diesel or a blend of diesel and natural gas. Saskatchewan Research Council photo *circa*. **1988.**

---

[45] One of these tractors was later converted to operate on ethanol fuel, see Chapter 6.

Figure 5.26. Photo of a Geo Metro natural gas conversion connected to a demonstration home fuelling station. Saskatchewan Research Council photo *circa.* **1996.**

Fermentation Technologies. Fermentation R&D at SRC had its beginnings in the early 1980s. By the early 1990s, there was significant commercial interest in Saskatchewan in the potential to produce ethanol for use as a fuel, through the microbial fermentation of the sugars in plants (starch and cellulose). SRC had ramped up its work with governments and regional businesses in evaluating, developing, and supporting ethanol fuel manufacturing pilot plants based on crops like wheat and alternative cellulose materials like straw (hence the term bio-ethanol) [64,74,77]. In 1997/98 SRC constructed a fermentation pilot plant and associated R&D laboratories at Innovation Place, Saskatoon, to enable practical work on potential new products including ethanol, crop inoculants, herbicides, pesticides, fertilizers, and animal vaccines [87].

Industrial Technology Transfer. SRC continued to provide "hands-on," "shop floor" technical services to a wide range of clients, particularly those in the manufacturing sector [71]. These services included SRC's delivery of NRC's Industrial Research Assistance Program (IRAP) in Saskatchewan, and also a new program aimed at inventors. The Inventor Services Program, which became better known externally as *"SRC's*

Inventors' Program," was launched in 1985 to help inventors with early-stage evaluations of their ideas.

SRC also continued to launch new periodical publications including the "*The Reliable Source,*" a quarterly "updates" newsletter for SRC clients that was produced from 1982 through 1983, the "*SRC Geoscience Newsletter,*" an annual newsletter specific to geoscience activities and advances that was produced from 1984 through 1986 (Figure 5.27), and a "*Remote Sensing Newsletter,*" that was produced quarterly from 1985 through 1986.

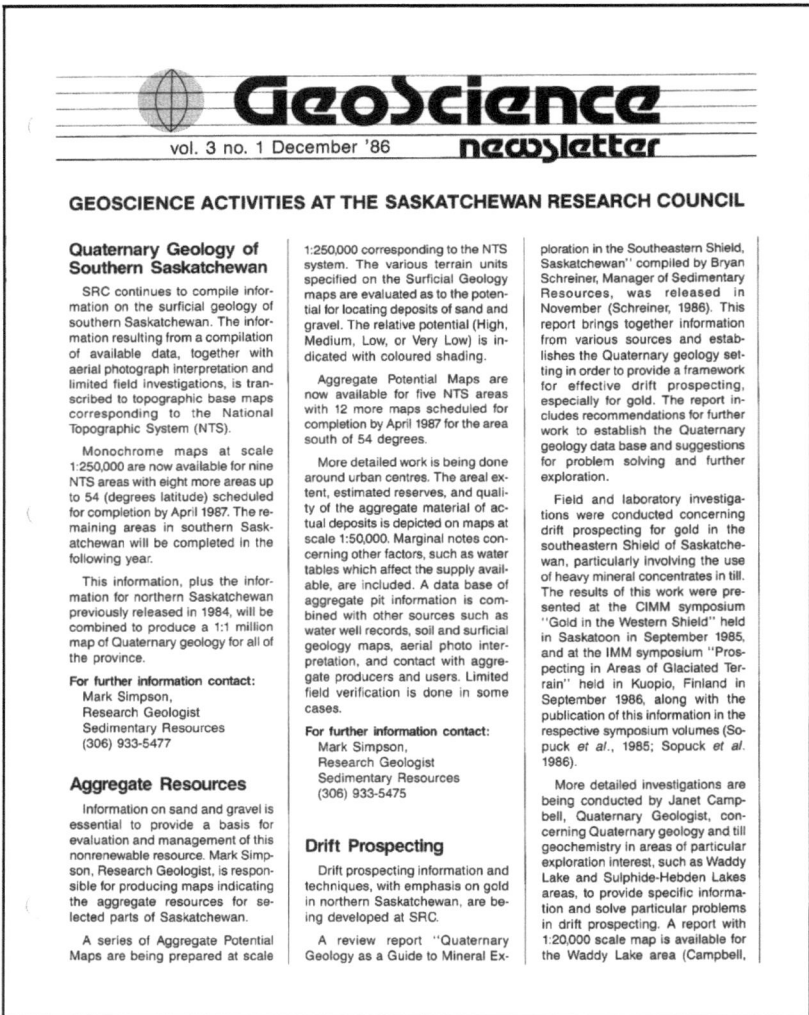

Figure 5.27. One of SRC's several newsletters of *The Commercial Years* era.

In 1994 SRC received a donation of $500,000 from Ian and Pearl Wahn enabling SRC to establish a perpetual *"Technology in Action Fund"* in memory of their Saskatchewan pioneer families [65,88]. Although this fund has been used in different ways over the years, in the 1990s it was mainly used to fund small projects aimed at assisting small- and medium-sized enterprises (SMEs), farmers, and rural communities [82].

Analytical Services. SRC had been conducting radon detection and analyses for industrial plants since 1979, and this program was expanded to radon analysis testing services for homes and buildings in 1989 [68]. By 1990, SRC was recognized as having one of the largest and best equipped analytical- and radiochemistry laboratories in Canada [79]. The analytical and the product research laboratories were the first of SRC's laboratories to become accredited to national quality standards (initially with the Canadian General Standards Board [65]). By the mid-1990s SRC's analytical laboratories were providing a full suite of analytical services spanning, air, water, plant, soil, and rock samples, with a large emphasis on environmental and radiochemical analyses [73].

Early in 1995 SRC hired Mr. Ron Woodward as SRC's fourth Permanent Head [64,73]. He served in this capacity, as President and CEO of SRC, until 1998 [78] (See Figure 5.28 and Appendix 9.4). By this time client contract work represented the majority of SRC's total revenues. The strategic plan was revised and the organizational structure changed to focus SRC even more strongly on conducting contract work for industry and, in particular, to focus on industrial client needs and client satisfaction [64,78]. By 1998 SRC's self-view had become that of a *"commercial entity"* and was explicitly focusing on becoming *"more competitive"*[46] [87].

SRC's TecMark technology commercialization program was created in 1995 and began assisting small companies with product commercialization [64]. TecMark International Commercialization Inc. was incorporated as a subsidiary of SRC in 1996, to assist companies with technology licensing and commercialization [87]. To assist with such activities, the Research Council Act was again revised in 2000 to enable SRC (or its subsidiaries) to acquire shares and other securities of other firms [23].

---

[46] This would shortly lead to issues with some businesses and ultimately with SRC's owner.

Figure 5.28 SRC's fourth Permanent Head, Mr. Ron Woodward took office as President and CEO in 1995 [64].

In 1998 SRC entered into a strategic alliance with Natural Resources Canada (NRCan), Saskatchewan Industry and Resources (SIR), and the University of Regina (U of R) to create the Petroleum Technology Research Centre[47] (PTRC) in Regina. Through this partnership, SRC and petroleum engineering faculty and students from the U of R began collaborating on an industry-led R&D program focused on high priority, common problems related to the production and recovery of Western Canada's petroleum resource.

---

[47] PTRC was incorporated as a not-for-profit company with the four partners (NRCan, SIR, U of R, and SRC) as Founding Members.

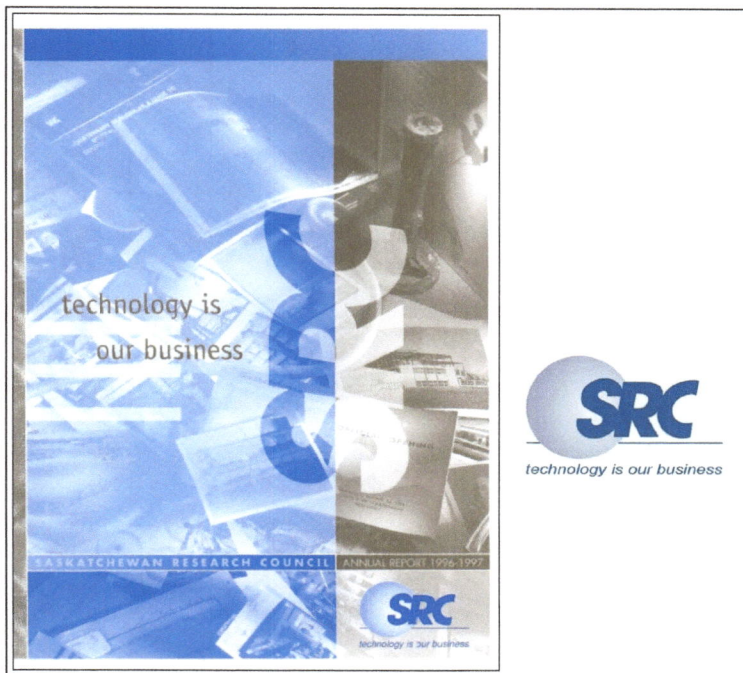

Figure 5.29. In 1995 SRC updated its company logo in concert with its revised strategy and intent to market services globally [64].

Figure 5.30. The Petroleum Technology Research Centre, Regina, in 2000, home for a new strategic alliance for SRC.

Early in 1999, Mr. Jim Hutchinson was promoted to become SRC's fifth Permanent Head [78] (Figure 5.31; see also Appendix 9.4). He served in this capacity, as President and CEO of SRC, until the following year (2000) [89], at which time the government dismissed the entire Board and replaced it with an interim Board. This created an organizational crisis that was hotly discussed in the media and in the Provincial Legislature [90-92].

Figure 5.31 SRC's fifth Permanent Head, Mr. Jim Hutchinson took office as President and CEO in 1999 [78,89].

The media reports referenced disagreements over how the partnership in PTRC should be managed but the government comments tended to focus on disagreement over strategic direction. For example, in the Provincial Legislature Minister Sonntag (the Minister responsible for SRC at the time) explained that "... *largely it came down to a fundamental policy difference or difference of opinion, and we ... simply couldn't find common ground, and as a result found this unfortunately as the only solution to the problem"* and when further questioned about the role of the government versus the role of the Board he explained that *"... it's the role of the minister and the government to generally lay out policy ..."* [92]. When questioned about the first role of the new, interim Board, Minister Sonntag responded that *"their first order of business was to dismiss the past CEO"* [92].

In January of 2001, the interim Board was replaced with the first seven members of a new Board of Directors, whose first major task was to recruit a new President and CEO.

Table 5.1 illustrates some of the changes experienced by SRC during *The Commercial Years*, 1983 – 2000. With staffing levels remaining at the 200-employee level, a more comprehensive approach to human resources management evolved in the mid-1990s bringing focus to such issues as externally competitive compensation systems, employment equity, employee wellness, and training needs assessments [73]. Similarly, increasing attention was being paid to SRC's occupational health and safety program (OH&S), and in the mid-1990s SRC adopted the philosophy that *"management of OH&S objectives has equal importance with the other primary objectives of the organization"* [73]. In 1998 a formal safety policy was developed and implemented, focused on providing a safe and healthy workplace. A common challenge for research and technology organizations (RTOs) is the need to balance the R&D needs of their government owner (and/or base funder) with the R&D and technological assistance needs of the industry that they also exist to serve [15,22,93]. Such balancing for SRC become one of the early imperatives for *"The New Millennium Years,"* but it would not be the only one.

Table 5.1. Illustration of some of the changes experienced by SRC during *The Commercial Years*, **1983 – 2000. See also Tables 3.1, 4.1, 6.1.**

	1983	2000
Total Revenue	$10,102 k	$19,166 k
Provincial Investment	$3,519 k	$7,947 k
Contracts	$6,282 k	$10,679 k
Fixed Assets (mostly laboratory equipment and fixtures)	$1,673 k	$6,927 k
Grants (to universities)		
Grant Expense	$118 k	$0 k
Grants as % of Expenses	~ 1%	0%
Employees	212	207
Reference	[34]	[82]

LAURIER L. SCHRAMM

6 THE NEW MILLENNIUM YEARS, 2001 - 2017

Late in 2001 SRC hired Dr. Laurier L. Schramm as SRC's sixth Permanent Head [94]. He has served in this capacity, as President and CEO of SRC, until the present day (See Appendix 9.4).

The initial priorities were very clear. In the aftermath of the events of 2000, the provincial government had quietly decided to launch a "mandate and mission" review of whether to keep SRC or not. Even if the government could be persuaded to keep SRC and maintain all or even most of its base funding support, SRC had become an outcast within most of the provincial government family placing a severe strain on the company's ability to negotiate and conduct work on behalf of the government. SRC's relationship within PTRC at the time was precarious at best. Meanwhile, morale within the company was at an all-time low. Notwithstanding these issues, the new Board was naturally very keen to have these issues resolved and the company put back on a growth trajectory. At this point in SRC's history there were thus three strategic imperatives: (1) convince the provincial government to keep SRC and maintain all of at least most of its base funding support, (2) repair and/or rebuild SRC's partnership relations within government and with PTRC, and (3) get SRC moving onwards and upwards again.

Most of the stories related to this part of the journey are beyond the scope of this book, but two of them have a direct bearing on everything that came next: gaining licence to continue to operate and be funded by the provincial government, and recalibrating and rejuvenating the organization itself. As it turned out the answer to the former provided the basis for the strategy for the latter.

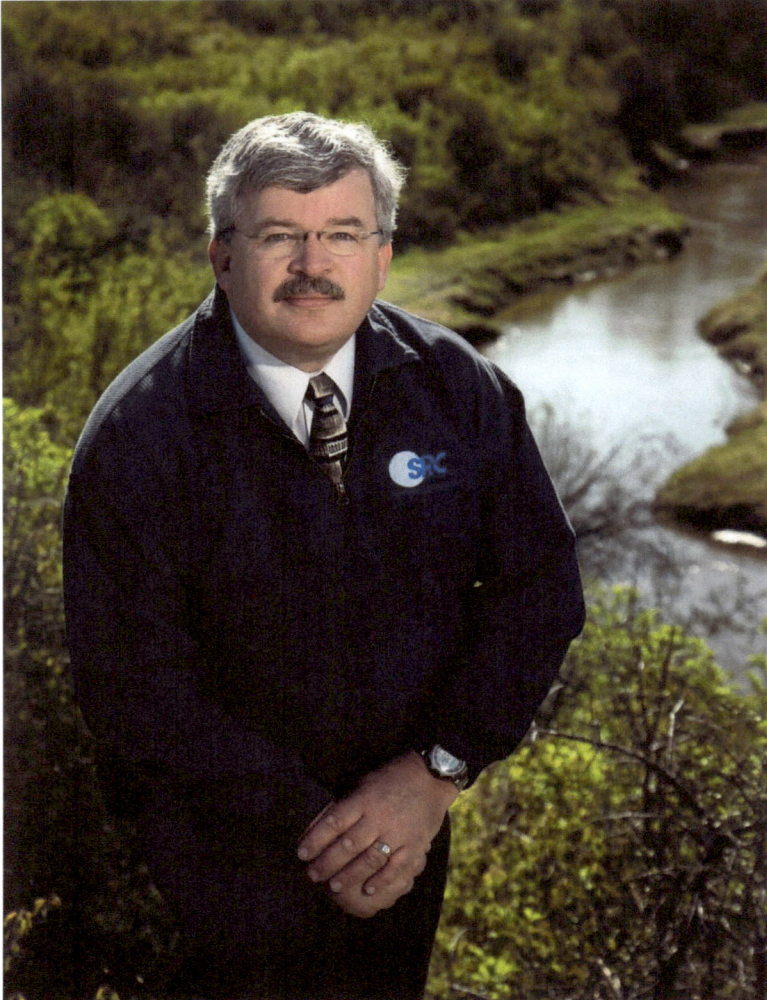

Figure 6.1. SRC's sixth Permanent Head, Dr. Laurier L. Schramm took office as President and CEO in 2001. Saskatchewan Research Council photo.

The first key question for SRC to address was its answer to the question *"Why should the government continue to support SRC?"* The government was already familiar with SRC's prior history and accomplishments so the key question became *"What should be SRC's go-forward value proposition to government?"* The essence of SRC's presentation to the Provincial Government in late 2001, was that SRC's Mandate and Mission were still valid and appropriate, that SRC fills a unique and critical position in the innovation continuum (see below), that SRC would continue to focus on helping strengthen and expand the Saskatchewan economy, and that SRC would commit to delivering demonstrable annual economic impacts to the province that would greatly exceed the annual base funding support provided by government. SRC's CEO projected that SRC should ultimately be able to leverage every dollar of annual base support with about another two dollars annually from industry, and with the work done with those funds achieve annual, incremental economic impacts of at least ten times the base funding amount. In the modern language of economic impacts that would mean achieving a Mandate Effectiveness Ratio (annual incremental economic impact divided by annual base funding from government) of at least 9.5 every year[48]. Thus, SRC's value proposition to government took the form of a return on investment commitment. The message received back from government was, to borrow a modern phrase, *"make it so."*

In order to recalibrate and rejuvenate SRC, it was decided to begin with a new strategy. The 2002 – 2007 strategic business plan was developed and implemented to focus on continuing to using science, engineering, and technology to serve industrial client needs, aligning with the provincial government's priorities and needs, and with a focus on client satisfaction (meaning all clients public and private) [95]. The largest initiatives under this strategy were to:

- o becoming increasingly entrepreneurial and growing the company in "breadth and depth," including broadening the client base to include more work for communities and for government,
- o diversify along the "innovation continuum," in order to manage a continuum of R&D activities that would connect all the way from new knowledge and new technologies, through applied research, pilot testing, development engineering and demonstration, and into the commercial realm of testing, analyses, and commercialization of technologies, and
- o increasing SRC's visibility and profile.

---

[48] This would be a huge stretch-goal for SRC. ARC, the leading Canadian RTO that was measuring and reporting economic impacts at the time, was achieving mandate effectiveness values averaging about 5 between 1995/96 and 1998/99; see also Chapter 7.

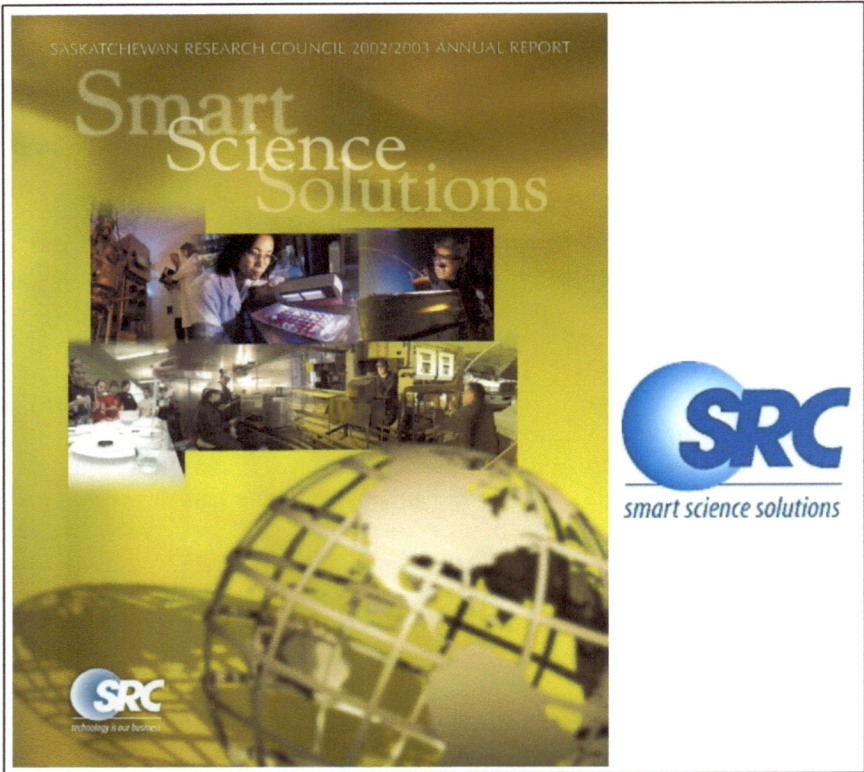

Figure 6.2. In 2003 SRC updated its company logo in concert with its revised strategy and intent to provide a balance of solutions to industry and government [96].

Two of the new features above involved recognizing government and communities as prospective major partners and major clients. For example, beginning in 2002 SRC created and managed an Office of Energy Conservation for the provincial government (see OEC, below), and in 2006 SRC landed a contract with the provincial government to manage the remediation of 37 abandoned cold war-era uranium mine and mill sites in northern Saskatchewan (see Project CLEANS, below). Also in the early 2000s, SRC began to work on major projects with northern and Aboriginal communities, sometimes as clients but more often as partners. For example, in 2003 SRC and Mistawasis First Nation launched a long-term collaboration on projects, beginning with one on groundwater resource evaluation and development [96], in 2006 SRC and the Federation of Saskatchewan Indian Nations[49] (FSIN) launched a collaborative project aimed at assessing water quality in First Nations communities [97], and in 2010 SRC and Cowessess First Nation partnered to develop and demonstrate a commercial-scale high-level wind energy generation and storage system [98]. The SRC/Cowessess demonstration project has been so successful that it has continued to meet the power needs of the local community while providing a modest revenue stream from selling surplus power into the provincial energy grid (Figure 6.3).

The 2007 – 2012 strategic business plan was aimed at continuing momentum along the directions of the previous five years, while striving to take on larger projects, diversify SRC's lines of business in strategic economic sectors and also along the innovation continuum, and continue to achieve exceptional economic impacts [99].

In this era, SRC's organizational structure was designed with two main groupings: a customer/sector-based grouping for the mostly external-facing Research & Technology (R&T) Divisions, and a common purpose-based grouping for the mostly internal-facing Corporate Divisions. The R&T Divisions were the ones conducting most of the applied research, development, pilot testing, development engineering and demonstration, and testing, analyses, and commercialization of technologies. As such the philosophy was that a customer/sector-based grouping would help these Divisions to maintain focus on the markets and on the clients, and their views on quality, while promoting client/sector visibility, cross-function communication, multidisciplinary teams, and *"market pull"*[50].

The Corporate Divisions were mostly charged with providing internal leadership and services to all parts of SRC, and as such the philosophy was that a common purpose grouping would help these Divisions to maintain focus on managing the SRC brand and those things that should be centralized, including aspects of governance, risk, brand management, safety, facilities and resources, and all-around effectiveness, while working to identify and remove barriers to growth, and generally provide enabling support to the R&T Divisions. With updates and changes from time to time, this hybrid of pillar and matrix structure has remained in place for the entire *New Millennium Years* era.

---

[49] Now the Federation of Sovereign Indigenous Nations.
[50] As opposed to *"technology push."*

Figure 6.3. The SRC/Cowessess First Nation demonstration of a commercial-scale high-level wind energy generation (Upper-Right) and battery storage system (Below-Left) in 2013. Saskatchewan Research Council photos.

In order to illustrate SRC's intended positioning in the innovation world, the innovation continuum concept was used, as shown in Figure 6.4. The innovation continuum, or *"technology S-curve,"* as it is sometimes called, follows a linear model for the development of technological innovations. Technological innovations actually don't usually evolve along a linear pathway like this, but the model is still used for illustration because of its simplicity and the fact that the developmental stages referred to are still legitimate and distinct.

In Figure 6.4 the green-shaded region labelled "Concepts/Fundamental" refers to discovery research, or *"basic research,"* as it is sometimes called, and the blue-shaded ellipse is drawn to only slightly overlap this area to indicate that this was not an area of SRC focus but rather an area from which SRC would draw new knowledge and understandings by collaborating and/or partnering with academic institutions.

The red-shaded region labelled "Feasibility/Development" refers to applied research, development engineering, and proof-of-concept at the laboratory bench-scale or *"research and development,"* as it is sometimes called, and the blue-shaded ellipse is drawn to fully overlap this area to indicate that this was a long-standing area of SRC focus that would be continued.

The grey-shaded region labelled "Piloting/Demonstrations" refers to pilot testing, scale-up engineering, and full-scale field/plant demonstration or *"piloting and demonstration,"* as it is sometimes called, and the blue-shaded ellipse is drawn to fully overlap this area to indicate that this was a somewhat new area of SRC focus that would be strengthened.

The yellow-shaded region labelled "Commercial Innovation" refers to commercial-sector activities, or *"products and services deployed in the marketplace."* The blue-shaded ellipse is drawn to only slightly overlap this area to indicate partly that only some SRC activities would be focused in this area (i.e., commercial testing and analyses) and partly that this is an area in which SRC would collaborate and/or partner with industry to assist with technology transfer and commercial deployment.

The Figure 6.4 illustration helped, not only by showing areas of high or low focus but also in identifying key interfaces (the regions where the blue ellipse only partly overlaps into the next region) where SRC would have to carefully manage collaborations and partnerships but being careful of real or perceived competition concerns on the part of external stakeholders.

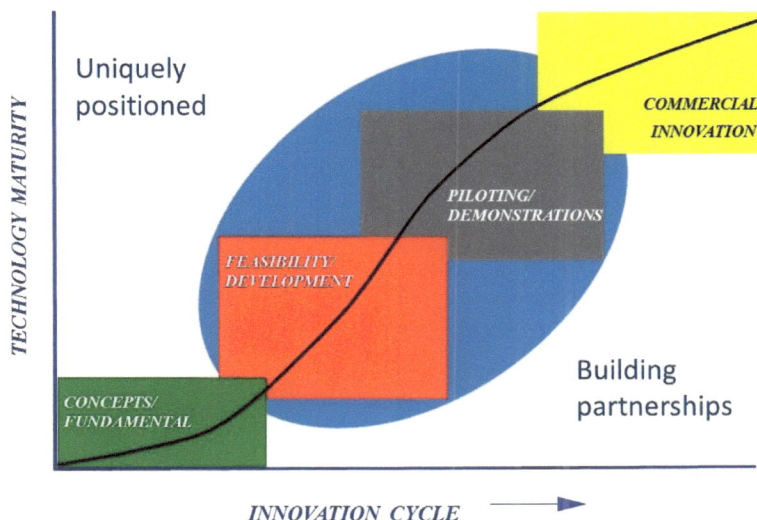

Figure 6.4. SRC's *New Millennium Years* areas of focus (in the blue ellipse) and positioning along the *"Innovation Continuum."*

Taking into consideration the two main organization groupings and the innovation continuum positioning, the Divisions and Business Units in the company were established and organized to be consistent with the province's strategic economic sector priorities while being sensitive to changing market conditions that would sometimes enable strong growth while at other times require substantial retrenchment. Overall the main areas pursued in this era were:

o   ***Fossil Energy***, with a focus on enabling the petroleum industry to recover a higher percentage of Saskatchewan's light and heavy crude oil (through enhanced oil recovery, EOR), increase production and be more efficient, with secondary areas of focus on improving transportation, upgrading of heavy oils into higher value products, and (market conditions permitting) to assist with the development of Saskatchewan oil sands, oil shale, and coal conversion.

o   ***Mining and Minerals***, with a focus on assisting the mineral industry with mineral exploration (especially for uranium and diamonds), process development, and transportation (especially pipeline transportation), with a secondary focus on tailings management.

o   ***Agriculture and Biotechnology***, with a focus on three areas: agri-value-added process development with and for feedstock, food, and other agri-product manufacturers, biotechnology process development (in such areas as biological fertilizers, crop protection agents, vaccines, bioethanol and other biofuels), and providing genetics-based services to help grow the agriculture and biotechnology industries generally.

o   ***Environment***, although not an industrial sector as such, SRC has long maintained a focus on helping stakeholders from all sectors with technological assistance in environmental aspects of water, air, and soil (especially in the regulatory compliance and sustainability aspects) and in understanding and adapting to the emerging impacts of climate change in Saskatchewan.

Although the above four represent the core areas of focus for SRC in this era, there have been periods of time in which other opportunities and/or other sectors have been pursued and developed. Examples of these are given further below.

Following a brief period of repositioning, SRC soon recovered the breadth and depth of project activities and revenues that had been built-up in *The Commercial Years* era and both provincial investment and client revenue numbers began to grow beyond those levels, and at an even more rapid pace than during the previous era (Figure 6.5). The number of employees grew commensurately, reaching 200 employees by 2002/03, 300 employees by 2005/06, and then more or less plateaued in the range 350 to 400 from 2008/09 onward.

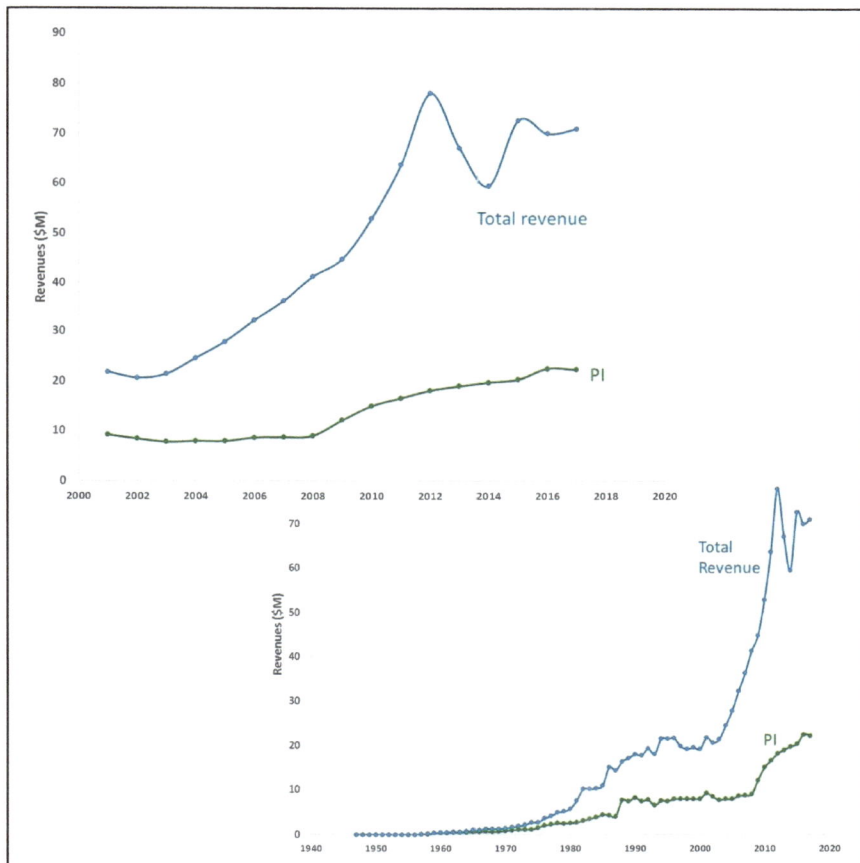

Figure 6.5. SRC revenue growth (total revenue and the provincial investment component, PI) in the *New Millennium Years* (upper) and over the full history to date (lower). The total revenue data points for the year 2016/17 are estimates.

As with previous eras, rapid growth placed stress on SRC's facilities and a series of minor expansions were implemented in Saskatoon (in the Galleria Building at Innovation Place and at the Northstar Centre), in Regina (in the PTRC building), and new space was acquired in Prince Albert (see below) and in Uranium City (see below). Ultimately, major renovations and consolidation of space became necessary in the Galleria, Atrium, and Downey Road Buildings in Innovation Place Saskatoon and "*The Big Move*," a $40 million four-year initiative was launched in 2015.

Taking into consideration the two main organizational groupings and the R&T Divisions positioning just described, the Corporate Divisions and their Sections were organized to be consistent with a growing set of internal leadership and service needs, particularly as related to those things that should be centralized. For example, the Executive Team initiated a drive in 2004 to take SRC's risk management practices to the Enterprise Risk Management (ERM) level, based on the AUS/NZ 4360 model, from there to more systematic risk assessment and reporting at the senior management and Board levels, and then moving towards ISO 31000. As part of this program, the Corporate Vice-Presidents and their teams were elevated to critical leadership roles in risk management across the whole company.

Safety has been important at SRC ever since it began to have employees in 1953, but rapid growth, and in particular expanding SRC's activities into larger-scale operations such as pilot plants and commercial demonstrations, brought new safety challenges. Accordingly, the degree of organizational focus on safety, which had already been increased between 2001 and 2004, was increased even further in 2004 when occupational health and safety (OH&S) was made SRC's over-riding priority. This was meant to ensure that safety never took a "back-seat" to business or other imperatives, and was the cornerstone of a new safety strategy, to build a "world-class" safety culture and record at SRC that continues to the present day[51].

An example of SRC taking a more active role in the community occurred in 2004 when SRC partnered with the City of Saskatoon's first responders (police, fire, ambulance, and emergency measures services) to run a full-scale, multi-agency disaster response exercise code-named BrickER, which involved 300 SRC and emergency personnel in a simulated tornado strike and building collapse event (see Figure 6.6) [100].

---

[51] In 2016 SRC was awarded the first large-company *Mission: Zero Award* from Safe Saskatchewan for "sustained improvement in their injury rate over time, and [working] towards transforming their organization culture and positioning injury prevention as a core value."

Figure 6.6. SRC helped lead a multi-agency mock tornado disaster exercise in 2004 to practice internal response procedures and inter-agency coordination [100].

Somewhat like safety, quality has been important at SRC ever since it began to conduct contract R&D for clients in 1956, but growing demands from the marketplace for independent accreditation of quality programs led to elevating all commercial laboratory operations (beyond Environmental Analytical Laboratories, which had moved to this some years earlier) to ISO 17025 accreditation between 2001 and 2004 (Petroleum Analytical Laboratories in 2001, and Geoanalytical Laboratories in 2004). Also in 2004, an overall SRC quality management strategy was developed and implemented. In 2014 SRC initiated a drive to take accredited quality management practices to company-wide status and ISO 9001:2008 accreditation for SRC was achieved in 2016.

Similar advances were developed to build on previous SRC systems in such areas as Human Resources Management, Diversity, Facilities Management, Information Technologies, Business Intelligence, Communications and Marketing, Legal, Major Projects, Financial Services, Purchasing, and Corporate Social Responsibility.

In order to more systematically maintain and advance technical excellence and internal innovation, several initiatives were developed and launched in the 2000s. A Technical Excellence Strategy was launched in 2007, together with an "*SRC Innovation Fund.*" The Innovation Fund process involves holding an annual internal competition for projects aimed at developing, strengthening, or diversifying new business initiatives, and in recent years has grown to an annual funding level of $1M/year. This represents a critical investment in the development of new products, processes, and services that are in turn devoted to SRC's core Mission and purpose. To the Innovation Fund has been added an annual "*SRC Innovation Forum,*" first launched in 2012, which is aimed at developing and nurturing a healthy internal "technology development pipeline" with substantial one-company awareness and interconnectedness. The *SRC Innovation Fund* and the *SRC Innovation Forum* are closely intertwined as the latter promotes and showcases the results of the former, while facilitating awareness, discussions, and the seeds of many new and innovative ideas that eventually become new *SRC Innovation Fund* projects. Out of these programs have come a stream of new products, process, and services that have been implemented internally, used in new client projects, and/or directly sold to clients in order to help them thrive and grow. Examples include SRC's HERC Power Systems [101] and SRC's High-Security Diamond-Sorting Gloveboxes (Figures 6.7 and 6.8).

Figure 6.7. The SRC HERC Power System, a rugged easily transported and deployed energy system (with integrated generator, battery, photovoltaic array, and inverter) equipped with remote control and monitoring systems. It can provide reliable off-grid power at reduced diesel fuel and maintenance costs compared with conventional power systems. The first commercial unit was installed at the Gunnar Mine Site (shown here) in 2015.

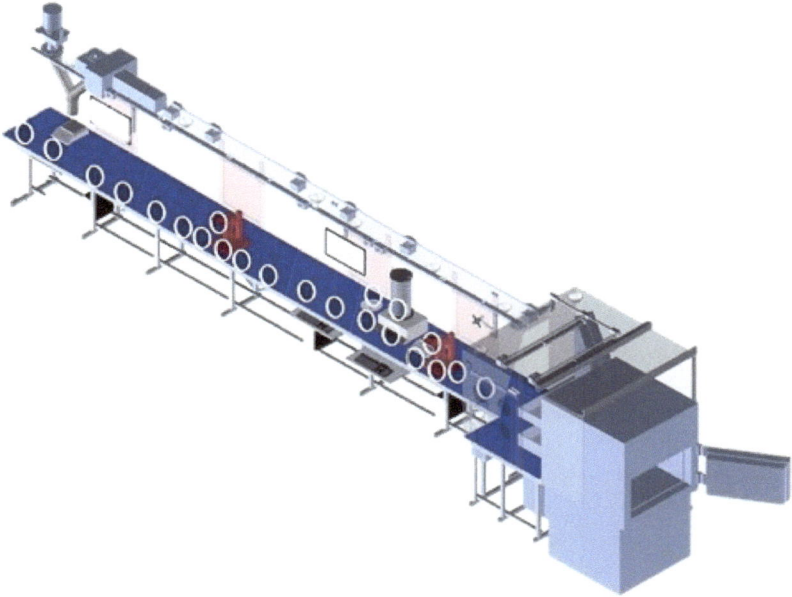

Figure 6.8. Illustration of an SRC high-security diamond-handling glove-box. This system, in different models, has been deployed internally and also sold commercially since its launch in 2015.

SRC continued to grow and develop its testing and analysis services during this era, both in order to support its strategic R&D activities and also to provide direct commercial services to clients. In the case of direct testing and analysis services for external clients, these were provided on the basis of unique and premium quality/service services offered on a nonsubsidized and premium-price for-profit basis (in order to avoid real or perceived unfair competition with the private sector) through SRC's:

- ○ *Environmental Analytical Laboratories*, which continued to expand and build on its unique "full-service" ISO 17025-accredited positioning, with a complete radiochemical analysis capability and SLOWPOKE nuclear reactor for uranium assaying and halogenated-organics analyses (for compounds like PCBs). Along the way, new techniques were developed, such as SRC's *Alpha Track Detector* for long-duration radon testing (Figure 6.10), and new services launched in such areas as radon and toxicity testing.

- ○ *Geoanalytical Laboratories*, also ISO 17025-accredited and which also continued to expand, but even more rapidly. The expanding demand for uranium analytical services for the exploration industry led to SRC building the largest uranium assay laboratory in the world (Figure 6.11). The next significant expansion was in analytical support programs for the diamond exploration industry. As industry demand grew, both from diamond exploration and development interests in Saskatchewan and around the world, SRC built and then successively expanded high-security diamond processing, extraction, and observation facilities encompassing (1) Kimberlite indicator mineral (KIM) assays, (2) micro-sized diamonds ("micros") from core drilling samples, and (3) macro-sized diamonds ("macros") from large-diameter drilling programs and/or from actual mining operations [98,100] (Figures 6.12, 6.13). These expansions led to SRC building one of the largest diamond assay laboratories in the world and becoming the preferred supplier for geoassay services to De Beers worldwide. Beginning in about 2006, in response to industry demand [106], SRC developed new assay methods for potash samples and established another new laboratory for this work, once again growing it to become the largest potash geoassay laboratory in the world. See Figure 6.13.

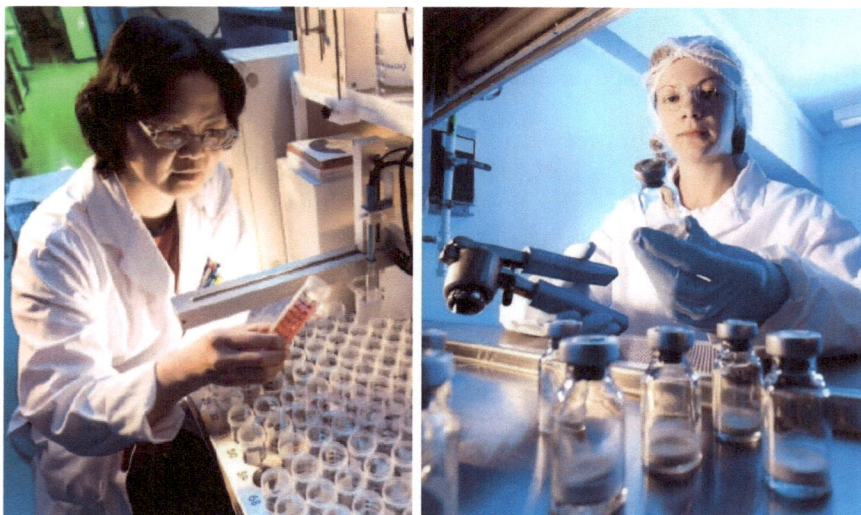

Figure 6.9. Analyzing for bacteria in 2005 (Left) and for quality-control in vaccine production in 2006 (Right). Saskatchewan Research Council photos [97,105].

Figure 6.10. SRC's Alpha Track radon monitor. Saskatchewan Research Council photo, 2008.

o ***Petroleum Analytical Laboratories***, also ISO 17025-accredited and which continued to offer transformer oil testing (including for PCBs), and petroleum industry testing and analytical services for a wide range of petroleum products, including natural gases, gas condensates, light hydrocarbons like gasoline, middle distillates like diesel, heavy crude components like asphalt and bitumen. In another expansion, SRC launched its *Biofuels Test Centre*™ in Regina, in 2006, to provide ethanol and biodiesel fuel testing services for producers and others [102]. See Figure 6.14.

Just as new and improved R&D and testing instruments, tools, and techniques were added to the laboratories just described, so too were they added across the R&T Divisions. Some of these are discussed further below, but an example is provided by the equipping and launch of a new *Advanced Microanalysis Centre*™ in 2009 [98,103,104]. This centre was built primarily to meet the needs of the mineral exploration industry, and was equipped with new tools like a state-of-the-art electron microprobe (used to identify and analyze the chemical composition of solid materials), and the entire facility was constructed to permit the handling and study of radioactive samples. In 2013 a QEMSCAN® instrument was added (Figure 6.15), which comprises a sophisticated electron microscope outfitted with multiple electron and X-ray detectors to enable the determination of the bulk mineralogy and liberation characteristics of uranium, potash, base metals, gold, rare earth, coal and other ore samples.

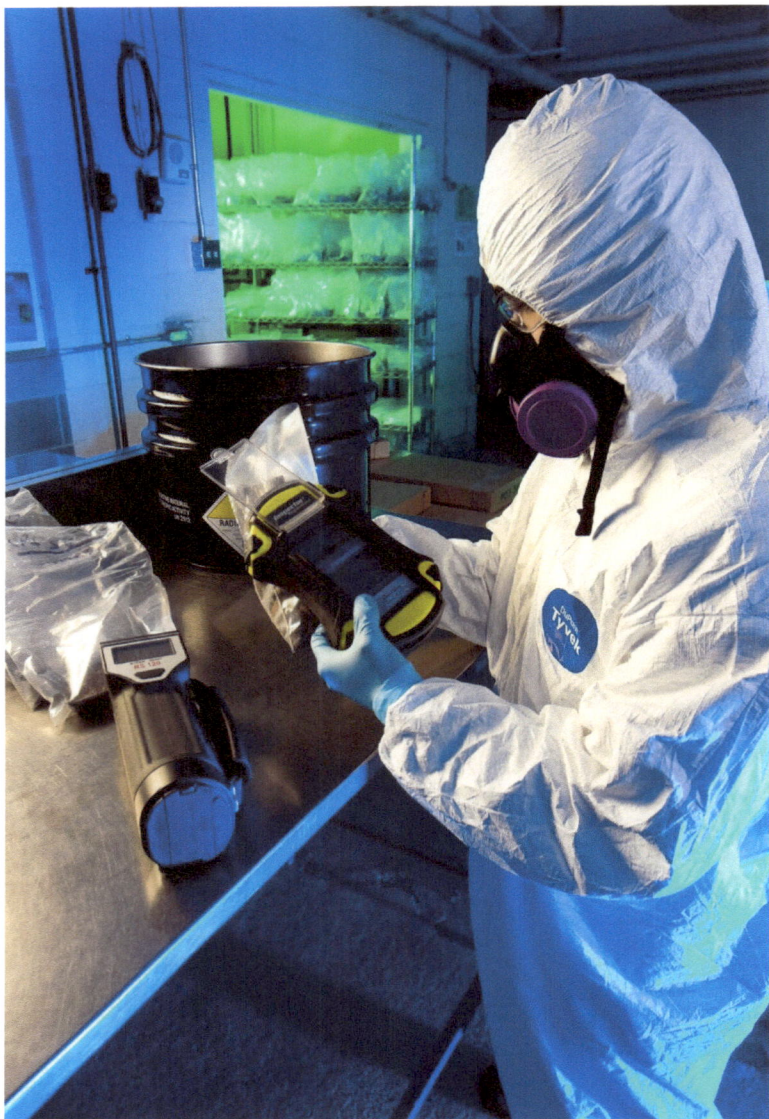

Figure 6.11. Checking incoming uranium samples for radioactivity in 2008. Saskatchewan Research Council photo.

Figure 6.12. SRC's dense media separation (DMS) plant for macro diamond separation in 2016. Saskatchewan Research Council photo.

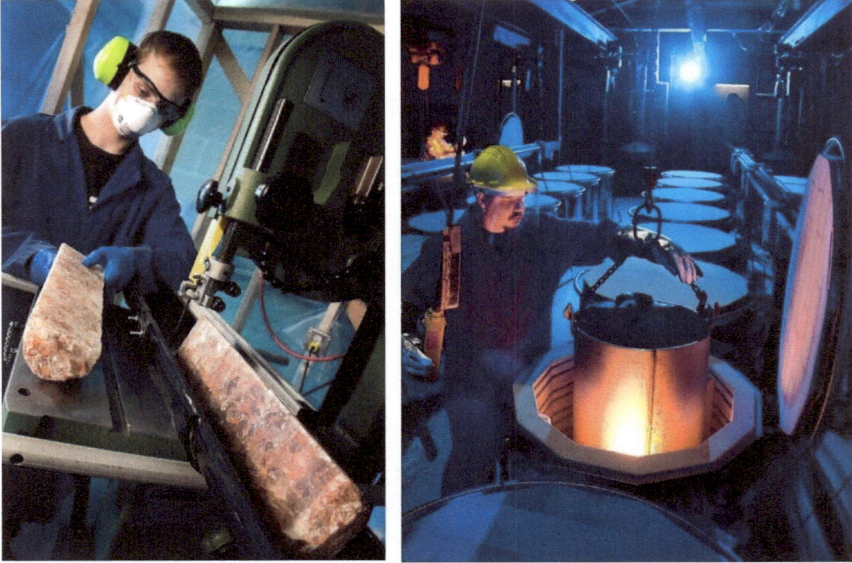

Figure 6.13. Preparing potash samples for geoassaying (Left) and kilns for diamond indicator mineral and/or micro diamond geoassaying (Right) in 2008 [106].

Some of the R&T program areas and accomplishments of *The New Millennium Years* will be highlighted in the following sections, which have been organized primarily in terms of strategic economic sector.

Agriculture and Biotechnology. Among the continuing programs in this area were work on the fermentation-based production of animal vaccines and plant inoculants, genetic testing of plants for desired traits and of animals for parentage and/or traits[52], and testing of samples for the presence or absence of genetically modified organisms (GMOs) [95,96,99,104]. Extensions of these programs addressed such diverse industry problems as helping the Saskatchewan Beekeepers Association develop mite-resistant honey bees [100], to helping the Canadian Wildlife Service manage the sustainability of elk herds [100], to helping the Canadian Forestry Service use genetic selection to develop fast-growing trees that could provide a biomass feedstock for energy production on a short-rotation basis [105].

---

[52] By 2013, market changes necessitated the sale of the animal genetics line of business (formerly part of the Bova-Can Joint Venture) and the Crop Evaluation (grain-testing) Laboratory; see Appendix 9.7.

In the late 2000s, SRC began developing genetic (DNA-based) tests for rapid identification of Canadian wheat and other grain varieties to support an export-oriented grain industry [106]. Most of the grain tests were ultimately licensed-out to commercial testing labs.

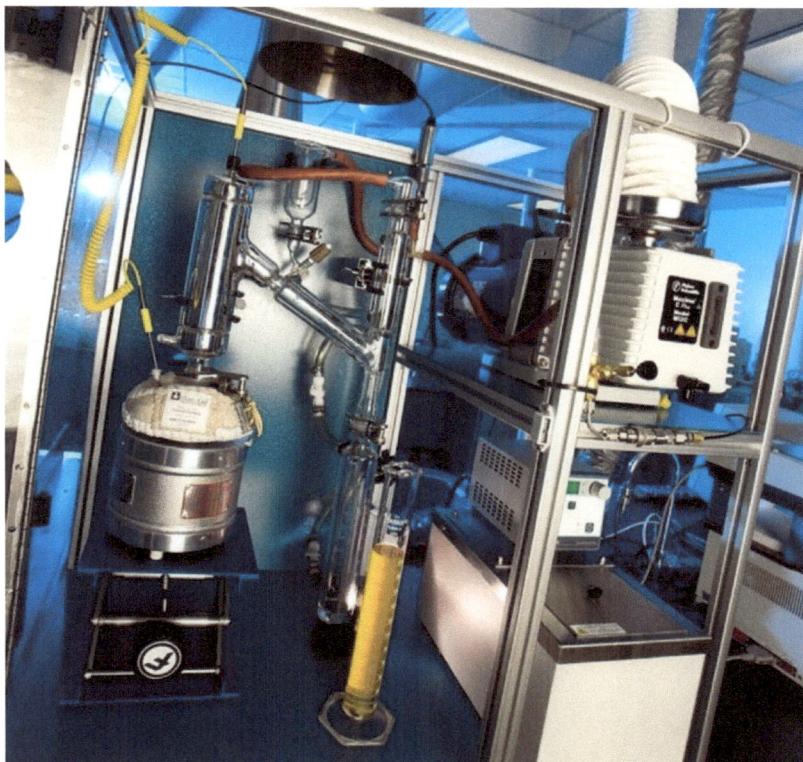

Figure 6.14. A vacuum distillation apparatus in the Petroleum Analytical Laboratories, in 2006. Saskatchewan Research Council photo.

With increasing government, community, and industry-level interests in bio-ethanol as a "green" fuel substitute, SRC carried out several programs aimed at supporting Saskatchewan's bioethanol and biodiesel producers by developing and/or adapting technologies for converting grains (like corn and canola) into ethanol, and providing ongoing technical support to the operating plants [95]. Other, more future-oriented programs were aimed at developing processes for converting waste cellulose materials (like straws, forest bark and branches, slough grasses, and the like) into ethanol and/or biodiesel [95]. Production testing and product verifications were provided by SRC's Biofuels Test Centre (mentioned above) and field testing in commercial farm machinery and transportation fleets was conducted under programs launched by the Office of Energy Conservation (OEC, see below) and continued by SRC after OEC was wound-up. In the biodiesel area, for example, a commercial demonstration program was conducted in 2009 through 2010 involving eight agriculture producers in the Foam Lake, Saskatchewan area, and encompassed 72 commonly used pieces of farm equipment including swathers, combines, trucks, and fuel-storage tanks. This program demonstrated the successful use of biodiesel fuel blends in farm equipment in real-world operating conditions, regardless of temperature, type of equipment, and long-term fuel storage (using biodiesel fuel produced in Saskatchewan, from canola) [104,107,108] (Figure 6.16).

Although the market for bioethanol and biodiesel did not expand to the degree many had hoped, SRC's work helped several Saskatchewan companies get started in this area, and developed and commercialized a number of technologies for the future.

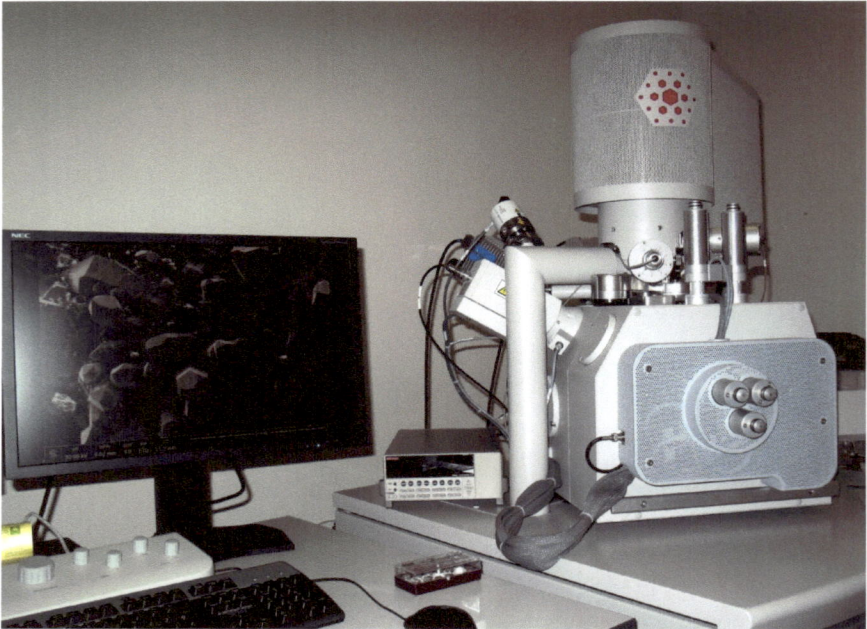

Figure 6.15. QEMSCAN® in 2013. Saskatchewan Research Council photo.

Figure 6.16. A farm tractor and bulk fuel tank (raised) from the biodiesel demonstration program near Foam Lake in 2010. Saskatchewan Research Council photo.

Numerous other related programs have been aimed at developing flexible, *"made in Saskatchewan,"* technologies to develop the province's huge, renewable waste and by-product biomass resources (ranging from slough grasses to forestry bark and branches, to sawdust). Most of these programs were aimed at developing conversion processes to produce syngas ($CO + H_2$) or pyrolysis oil, and processes for the conversion of syngas or pyrolysis oil into value-added products such as electricity, heat, fuel, and/or other value-added products. Examples range from developing a sawdust gasification (heat and power) process for a lumber mill in La Ronge, to developing catalytic processing technology for proposed bioethanol plants in Nipawin and elsewhere in Saskatchewan and the U.S., to developing a demonstration biodigester (making biogas from animal waste) for a permanent working display at the Canada Agriculture and Food Museum [96,109,110]. Building on the La Ronge gasification experience, SRC developed other approaches and demonstration projects in small-scale *"combined heat and power"* (CHP, also termed cogeneration) in which heat and power are produced simultaneously. In 2009, for example, a full-scale plant demonstration was launched at Inland Metals, in Regina, involving a natural gas CHP unit suitable for a single manufacturing plant [98,104]. Such technology is capable of large improvements in energy efficiency, proving a cost advantage to the industry with reduced greenhouse gas (GHG) emissions and reduced demand on the regional power grid.

SRC's fermentation laboratories and pilot plant (Figure 6.17), whose origins date back to 1998, have been focused in different areas as market needs have evolved (and sometimes cycled). Whereas in earlier eras these foci had included support for fermentation-based bioethanol production and/or the production of animal vaccines and plant inoculants, in the *New Millennium Years* SRC experienced reduced market-pull for work in these areas leading to a strategic shift to be able to serve broader markets. To enable this, in 2015 SRC partnered with the National Research Council of Canada (NRC) to bring the capabilities of both organizations together to work on bio-manufacturing processes and products for industries both in and beyond the traditional agriculture and health sectors (i.e., including energy and mining and minerals) [111,112]. By 2016, Alberta Innovates – Technology Futures and Bio|Food|Tech (in PEI) had informally joined into this cooperative effort as well.

Figure 6.17. SRC's Biosafety Level 2 Fermentation Pilot Plant in 2007. Saskatchewan Research Council photos.

Forestry and Forest Products. A continuing area of work for SRC has been assisting the forest products industry in traditional areas like process development and optimization work in pulp mills, and newer areas like biomass energy developments for sawmills, such as using sawdust, bark, and branches as feedstocks for bioenergy and heat production. Other continuing work has been in the areas like agroforestry and afforestation, particularly where the intention is to produce the trees as a product.

In 2009 SRC's forestry activities moved upstream with the acquisition of the program responsibilities of the former Forest First organization in Prince Albert, principally in the areas of agroforestry and bioenergy [113].

With this transition, SRC assumed responsibility for work in reforestation through Forest Trust arrangements in the province, and also for work in forestation involving research plots of land dedicated to evaluating new tree species, such as willow trees that can be grown in short rotation and used as a green, renewable feedstock for bioenergy production. With this acquisition, SRC also established a small, permanent presence in Prince Albert (Figure 6.18). To aid in the forestry R&D work, SRC established a collaborative agreement with the Conservation Learning Centre (located near Prince Albert) for boreal forest work at the field laboratory (Figures 6.19 and 6.20) and other programs at the latter's demonstration centre for soil and water conservation technologies [114].

Figure 6.18. In 2009 SRC acquired space in the Saskatchewan Forest Centre building in Prince Albert. Saskatchewan Research Council photo.

Figure 6.19. Shelterbelt trees at the Conservation Learning Centre's field laboratory near Prince Albert. Saskatchewan Research Council photo, 2010.

Manufacturing and Value Added Processing. SRC continued to work with industry, particularly small and medium-sized enterprises (SMEs), to enhance the ability of the manufacturing industry to compete globally. This was mostly done by assisting clients in the design, development, and evaluation of new products, processes, and services. A wide range of development engineering expertise and resources have been offered in this area, including mechanical design, rapid prototyping (Figure 6.21), instrumentation & electronics, intelligent control systems, and diagnostics.

Numerous tools, instruments, machines, and prototype products have been designed, developed, and manufactured in-house for clients large and small over the years. Sector examples range from energy (tank level sensors, oilfield communication systems, and alternative fuels), minerals (underground measurement and survey tools), agriculture (agricultural vehicles and machines), manufacturing (alternative fuel delivery and control systems and moulds for plastic component manufacturing), and environment (underground survey tools), to name just a few (see Figures 6.22 – 6.24).

Figure 6.20. SRC forest ecologist and Distinguished Scientist Dr. Mark Johnston[53] has been using Saskatchewan's Island Forests as outdoor "laboratories" for studying potential climate change impacts on important characteristics such as growth patterns and susceptibility to pests and forest fires [98].

[53] For a list of SRC's Distinguished Scientists and Researchers Emeriti see Appendix 9.5.

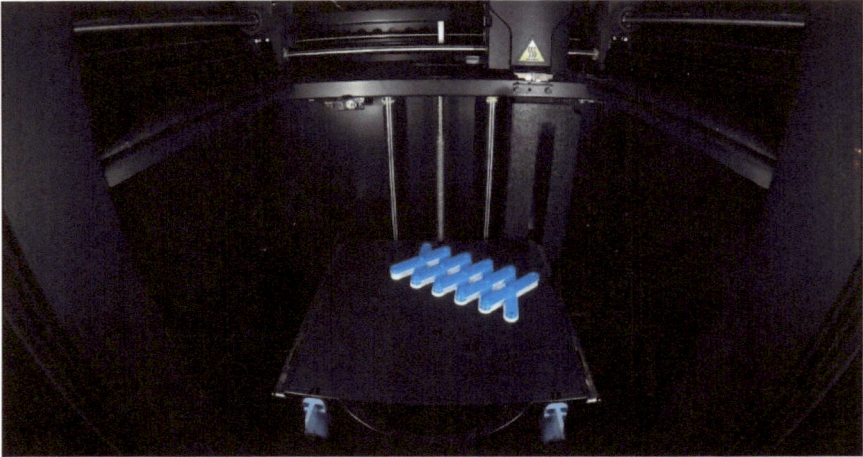

Figure 6.21. Fish-eye view inside an SRC 3D rapid prototyping machine. Saskatchewan Research Council photo.

Figure 6.22. A Borehole Autonomous Survey System (BHASS) deployed in an underground uranium mine in 2011. This is a rugged, semi-autonomous system that can be used for inertial, radiometric, and video probe surveys up vertical boreholes up to 60 metres in length. Such boreholes are accessible from underground tunnels at depths of approximately 500 metres. The upper-right and lower-left photos show views from each end of the system.

Figure 6.23. SRC's underground survey camera developed for use in hazardous underground mine cavities. It was first deployed for surveying abandoned underground mine cavities in 2015.

Figure 6.24. An example of municipal infrastructure innovation: SRC developed a *"Pipe Removal and Replacement System,"* shown here being lowered into an excavation hole in Regina, where it will be used to remove a lead water pipeline and replace it with a new copper pipeline [115].

Energy Efficiency. In the early 2000s both the provincial and the federal governments encouraged SRC to increase the breadth and depth of its work in energy efficiency (mostly in housing and buildings) and alternative energy (mostly in terms of developing alternative energy sources and in terms of alternative fuel vehicles). SRC's work in building performance received a boost and an expansion when the province asked SRC to create and manage a suite of energy efficiency and energy conservation programs. To this end, the Saskatchewan Office of Energy Conservation (OEC, Figure 6.25) was launched in 2002 [116] and operated through 2008. OEC's mandate was broader that just building energy efficiency, however, and OEC also became involved in energy efficiency programs generally. OEC supported such initiatives as the Model National Energy Code, R-2000 program, Energy Star programs, National Fleet Challenge, and the Municipal Energy Efficiency Program. Some of these programs achieved major energy reduction impacts, such as the programs with municipalities that focused on reducing power usage in swimming pools (heat and light) and curling and skating arenas (cooling, heating, and light) [100]. The results of these programs have been applied in other regions of Canada and at least as far away as Australia [105].

Figure 6.25. SRC launched the Office of Energy Conservation in 2003.

Among the OEC's accomplishments were full-scale, real-world demonstrations of energy efficient housing including the Factor 9 House (Figure 6.26) and affordable, ready-to-move manufactured homes that exceed national requirements for energy efficiency and meet the needs of Northern communities (developed in collaboration with the Agency Chiefs Tribal Council, Figure 6.27) [95,100,105].

Although the OEC was wound-up in 2008 due to diminished funding, SRC continued to advance other alternative energy programs including continuing to work with municipalities to adopt energy efficient infrastructure like lighting, and alternative energy sources like solar water heating [106], working with small businesses and the public to assist with solar- and wind power systems [98,106], and working with Canada Mortgage and Housing Corp. (CMHC) on home energy conservation technology monitoring and evaluation [115].

Figure 6.26. The SRC/OEC "Factor 9" demonstration home, which uses 90% less energy and 50% less water than typical homes built with 1980s technology. Saskatchewan Research Council photo, 2008.

Figure 6.27. The SRC/OEC affordable Northern housing demonstration house, which uses 35% less energy and 20% less water than the Canadian Model Energy Code for Houses but at no incremental cost over conventional construction. Saskatchewan Research Council photo, 2007.

Alternative Energy Vehicles. In the area of alternative fuel vehicles, SRC's previous work on natural gas vehicle conversions was broadened to other possible fuels. This led to the development of a series of demonstration vehicles that could operate on natural gas, ethanol, and/or hydrogen. A few of these were complete conversions to a single, but new fuel source, such as natural gas cars and an ethanol-fuelled tractor (Figure 6.28, Middle). The latter was one of the natural gas conversion tractors referred to in Chapter 4, but converted to operate on ethanol fuel, rather than 100% natural gas, and demonstrated in various parts of Saskatchewan and Alberta before being transferred to the Pound-Maker Feedlot & Ethanol Plant in Lanigan, Saskatchewan for commercial use (using their own ethanol fuel).

Many other vehicles were developed and demonstrated as after-market conversions, see Figure 6.28.

Figure 6.28. SRC's ethanol-fuelled vehicles (Middle photo) in 2006 [97], a range of alternative-fuel trucks (Upper photo) *circa* 2004, and a natural gas/gasoline hybrid electric SUV (Lower photo) in 2012. Saskatchewan Research Council photos.

Such vehicles were based on SRC's Dual-Fuel® intelligent systems concept, by which a vehicle could operate on its original fuel type or "on the fly" dynamically blended original plus alternative fuel (spanning the full range of 0 to 100% substitution) [117].

Examples of these were an ethanol/diesel truck (Figure 6.28, Middle), several natural gas/gasoline, hydrogen/diesel, and hydrogen/gasoline trucks (Figure 6.28, Upper), and a three-energy-source natural gas/gasoline hybrid-electric truck (Figure 6.28, Lower) [106,115]. (These included the world's first hydrogen Dual-Fuel vehicles to be developed and field demonstrated.) All of these alternative fuel vehicles demonstrated the ability of commercialization-ready technologies to improve fuel economies and reduce emissions, including greenhouse gas emissions.

SRC also launched Saskatchewan's first hydrogen loading and fuelling station, demonstrating hydrogen from capture to highway, in 2010. For this commercial demonstration by-product hydrogen was captured at a chemical plant, compressed and dried, then transported to a SaskEnergy fuel station. The hydrogen was used to fuel a demonstration fleet comprising seven pickup trucks (four SRC trucks and three SaskEnergy trucks) that had been adapted to Dual-Fuel® hydrogen operation by SRC [98,118] (see Figure 6.29).

Figure 6.29. A hydrogen fuel pump (left) and transportation trailers (below) at a SaskEnergy fuel station in 2010. Saskatchewan Research Council photos.

Mining and Minerals. SRC's work in mineral exploration programs continued to focus on uranium, diamonds, and potash. In 2005 a 3D virtual reality (VR, Figure 6.30) and modeling facility was opened and initially dedicated to helping uranium and potash mining companies design their exploration drilling programs with the ability to produce a holographic display representing multiple features, datasets, and models [97,100,105]. Within a decade, desktop 3D VR systems had evolved to the point where they could displace such large facilities, at least for SRC's applications, causing SRC to switch to the new smaller systems and the original large 3D VR system was donated to the University of Regina late in 2012. This illustrates how quickly technology can change, part of the reason SRC has had to reinvent its capabilities and services so frequently over the decades.

Mineral processing support programs continued to focus on uranium, diamonds, coal, and potash. In the early 2010s increased mineral exploration and development, particularly in rare earth minerals, led SRC to construct a new, larger and more flexible mineral processing pilot plant, which opened in 2013 [119,120]. The pilot plant supports companies involved in processing minerals such as potash, gold, base metals, diamonds, coal, oil sand, oil shale, and especially rare earth minerals (Figure 6.31).

Figure 6.30. 3D VR Centre in 2006 [105].

Figure 6.31. Part of SRC's new Mineral Processing Pilot Plant in 2013. Saskatchewan Research Council photo.

Fossil Energy. The PTRC issues from the *Commercial Years* were dealt with by a trifold strategy of:

1. Engaging as a full partner in advancing PTRC's Mission, which in 2001 was [121]: *"The PTRC will initiate and support research and development projects aimed at enhancing the production and recovery of Canadian petroleum resources by drawing primarily but not exclusively upon the expertise of the Petroleum Branch of the SRC and the Engineering Faculty of the University of Regina"*

2. Engaging with U of R researchers on core PTRC projects that lie mostly in the interface between fundamental and applied research (illustrated by the lower-left region in the innovation continuum shown in Figure 6.4), and

3. separately conducting contract projects for industry that involve a high degree of need to protect proprietary information and which lie in the development engineering, piloting, and field demonstration realms (illustrated by the upper-right region in the innovation continuum shown in Figure 6.4).

For example, SRC partnered with PTRC on the design and construction of an innovative three-dimensional high-pressure scale model to enable the physical simulation of field conditions as part of the development of enhanced oil recovery methods such as solvent-vapour extraction (SVX) [98,100]. Other early partnership successes with PTRC were the Viking program, which addressed conventional light oil production challenges in west-central Saskatchewan's Viking oil reservoirs [97], and the Joint Implementation of Vapour Extraction (JIVE) program that combined operating field pilot data with laboratory investigation and numerical simulation to improve the solvent vapour extraction process for heavy oil production [100]. Projects like these enabled the best knowledge, skills, and experience from an academic institution (University of Regina) and a research and technology organization (SRC) to play important roles in a collaborative program.

Figure 6.32. SRC's 3D high temperature, high-pressure physical model in 2009. Saskatchewan Research Council photo.

Among the continuing fossil energy programs of a more proprietary nature were programs aimed at enhancing the recovery of conventional light and heavy oils from Saskatchewan reservoirs, often by custom-designing and/or modifying reservoir-drive processes ("floods") based on the injection of gases, solvents, steam, alkali/surfactant/polymer solutions, or combinations thereof, including SRC's thermal solvent extraction (TSX) and solvent vapour extraction (SVX) processes. Some key field pilot tests and demonstrations include:

o The Plover Lake heavy oil field near Plover Lake, Saskatchewan, has been challenging for operators to produce. SRC assisted Nexen Inc. with technical support for the VAPEX (VAPour EXtraction) EOR process technology being test-demonstrated, and with real-time measurements and reporting from the site 24 hours a day, seven days a week [95,96].

o The Bakken formation, which underlies much of the Williston Basin, contains up to 500 billion barrels of conventional light oil. Despite the introduction of horizontal wells with multistage hydraulic fracturing, the wells' performance decline extremely rapidly due to reservoir connectivity issues. SRC's work in this area has included characterizing, screening, and evaluating EOR techniques for the Bakken, and has focused on developing customized waterflood and gas injection EOR process solutions [104,122].

o The Alberta Basin heavy oil deposits extend into Saskatchewan with over 2.3 billion barrels in-place on the Saskatchewan side in the "Viking Play" alone [123]. Cold heavy oil production with sand (CHOPS) enables some of the oil to be produced but leaves about 90% behind as unrecoverable. SRC's work in this area has focused on developing and demonstrating EOR processes that take advantage of the wormholes created in the reservoir by CHOPS (sometimes referred to as Post-CHOPS processes) [124].

In several other, more forward-looking programs SRC worked to advance processes that could be used to develop other Saskatchewan energy resources in environmentally sustainable ways. Some examples of this include:

o The design and construction of a dedicated 3D scaled-physical model for use in the development of a shallow *in situ* oil sand extraction process. This was done with substantial assistance from Oilsands Quest Inc. [106], with the initial target being a process such as SRC's thermal-solvent extraction (TSX) for Saskatchewan's oil sands at Axe Lake [98,99,125].

○ Resource and process evaluations for Saskatchewan's oil shales (in the Pasquia Hills region, near Hudson Bay) [126].
○ Process evaluations and development for coal-to-liquids and coal gasification technologies for Saskatchewan's coal resources.

Pipeflow Technology. Continuing programs were maintained in slurry hydrotransport work for the oil sands industry and others, while at the same time numerous other projects were initiated to support the continued development of Saskatchewan's conventional oil and gas and mining and mineral industries.

As the petroleum industry become more interested in the possibility of pipelining heavy oils and bitumens in Saskatchewan and Alberta, another partnership program with PTRC was launched to develop the technology needed for water-assisted pipeline transport of bitumen, heavy oils and co-produced sand [100]. Such pipeline operations could avoid the need for hot oil and diluent pipelines, reduce heavy truck and/or conveyor transport, and provide environmental sustainability benefits [100]. With federal and provincial government support the construction of a new high pressure – high temperature facility was announced in 2011 [104], and launched in 2015 [127]. This facility, the multi-purpose *Shook-Gillies High Pressure, High-Temperature Pipe Flow Facility*, is flexible enough to be able to handle a range of volatile, sometimes hazardous and/or explosive, transportation fluids – from solvents and diluents to heavy crude oils to diluted bitumen.

In *The New Millennium Years,* a combination of operational, legislative, and public pressures led to renewed efforts on the part of the mineral processing industries to improve and/or optimize their tailings management strategies. As a result, the Pipeflow Technology Centre has been involved in a number of tailings transportation and management projects, particularly in potash.

Environment. Several of SRC's long-standing geographic information system (GIS) related programs (like remote sensing and creating digital land-cover maps; see Chapter 5) continued but evolved into programs under the Saskatchewan Geospatial Imagery Collaborative (SGIC), which was created in 2008. SRC manages the SGIC, on behalf of this partnership among many organizations, that shares knowledge and costs relating to acquisition and use of satellite and aerial imagery for mutual and public benefit.

Figure 6.33. SRC chemical engineer, Distinguished Engineer, and Researcher Emeritus[54] Dr. Randy Gillies, in SRC's Pipeflow Technology Centre.

---

[54] For a list of SRC's Distinguished Scientists and Researchers Emeriti see Appendix 9.5.

SRC's long-standing groundwater and hydrogeology programs continued into the mid-2000s. In 2005, SRC transferred its provincial groundwater observation well network and related programs to the Saskatchewan Watershed Authority (SWA). In subsequent years SRC's water focus shifted to water quality issues of concern to Aboriginal and northern communities, such as SRC's groundwater resource evaluation and development program with Mistawasis First Nation mentioned above.

Air quality testing and monitoring programs continued and expanded in this era with SRC conducting almost all of the industrial stack emissions monitoring in the province plus (by 2016) most of the airshed monitoring across the most heavily populated, southern half of Saskatchewan (Figure 6.34). Here again, new instruments and techniques were deployed including airpointer® monitors and advanced hydrogen sulphide monitors. The airpointer® monitors (Figure 6.34) collect the data every minute and can remotely report it hourly, while covering a range of air pollutants (from nitrogen dioxide and ozone to hydrogen sulphide and sulphur dioxide) together with environmental conditions (from temperature and humidity to wind speeds and pressures) [104].

Figure 6.34. Environmental chemist Keith Wallace checks an airpointer® multi-species air quality monitor in 2010 (above), and SRC employees conduct emissions sampling near the top of a large industrial stack in 2016 (right). Saskatchewan Research Council photos.

SRC continued to maintain its long-standing "principal" Climate Reference Station (CRS) in Saskatoon, celebrating its 45-year anniversary and new instruments in 2009 [128]. Some examples of recorded Saskatoon weather extremes since 1963 include:

o   Highest temperature: 41.0°C on June 5, 1988,
o   Lowest temperature: -43.9°C on Jan. 22, 1966 and Jan. 29, 1969,
o   Most precipitation: 99.4 mm on June 24, 1983,
o   Earliest last spring frost: May 1, 1977, and
o   Latest last spring frost: June 14, 1969.

In response to demand for a comprehensive climate station in Central Saskatchewan, a second Climate Reference Station was launched at the Conservation Learning Centre (CLC) near Prince Albert in 2011 [129]. Both stations provide climate data to the agriculture and forest industries, insurance agencies, health, construction firms, media, hydrology sector, and government agencies, among others. The more northerly station is also in a key location for collecting climate data to assist communities and companies in making informed decisions about how to respond to climate change and climate hazards.

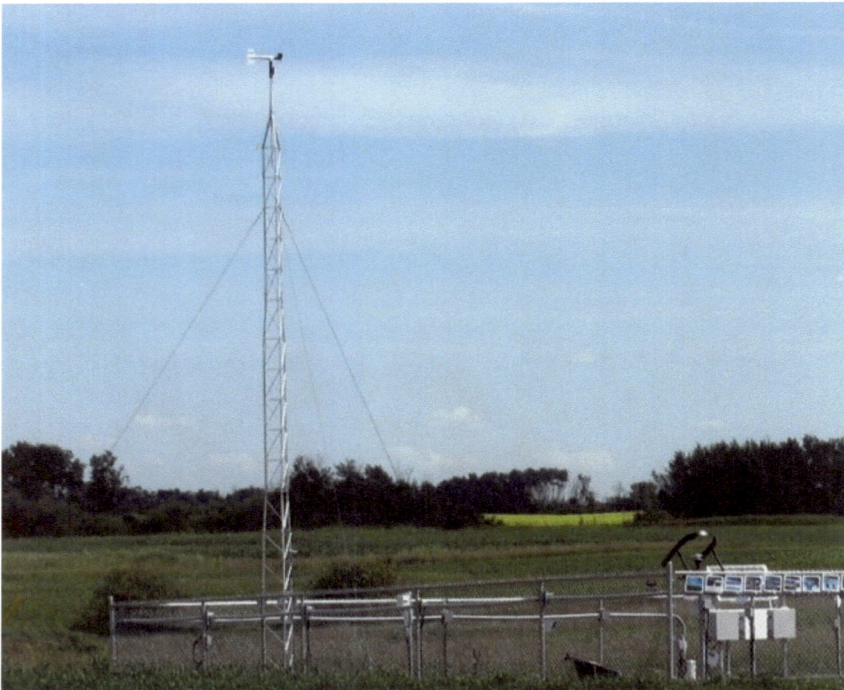

Figure 6.35. The Climate Reference Station at the Conservation Learning Centre near Prince Albert in 2011. Saskatchewan Research Council photo.

SRC's research into climatology and climate change continued in this era but made a strategic shift to more strongly emphasize methods for adapting to the impacts of climate change. When the international Kyoto Protocol on climate change came into force in 2005, government demand for research and development on measures to adapt to and mitigate climate change intensified. This was particularly the case for Saskatchewan, which NASA had identified as *"ground zero"* *"hotspot"* for North America [130,131], meaning that the greatest climate change impacts on the continent would hit this province, raising concerns regarding Saskatchewan's ecosystems, economy, and public health [100].

Figure 6.36. SRC climatologist, Distinguished Scientist, and Researcher Emeritus[55] Dr. Elaine Wheaton was a co-recipient, as part of IPCC, of the 2007 Nobel Peace Prize [106].

[55] For a list of SRC's Distinguished Scientists and Researchers Emeriti see Appendix 9.5.

SRC's climate change work had impacts beyond just Saskatchewan. For her contributions to the international climate change research effort, SRC climatologist Dr. Elaine Wheaton was a co-recipient, as part of the International Panel on Climate Change (IPCC), of the 2007 Nobel Peace Prize (shared with Al Gore) for building and disseminating knowledge about climate change – see Figure 6.36 (above).

Project CLEANS. In 2006 SRC was contracted by the Province of Saskatchewan to manage the remediation of the former Gunnar and Lorado uranium mine and mill sites, as well as 35 satellite uranium mine sites in the Uranium City area of Northern Saskatchewan [97,98,104,106]. All 37 sites were abandoned, orphaned legacies of the Cold War era [132-138] (see Figures 6.37 and 6.38). This work, which is still underway at the time of writing, involves assessing the sites to determine environmental impacts and developing remediation plans to remove or lessen those impacts, conducting remediation activities in a manner that meets or exceeds regulatory requirements, and ultimately monitoring the sites post-remediation, to demonstrate achievement of the final remediation objectives. In addition to environmental protection, this project also has substantial public safety objectives because, in addition to the obvious radiation safety aspects, many of these sites have hazards related to aging structures, openings to underground shafts, and chemical hazards. These have been described elsewhere [132-135,138].

Figure 6.37. Aerial view of the flooded open pit at Gunnar in 2006. Author Photo.

Figure 6.38. The Gunnar head-frame in 2006. Author photo.

The complexity of this project and the range of hidden hazards has required an adaptive management strategy, including aspects of R&D and new technology testing and demonstration as the work has progressed. The actual remediation work began in 2009, and at the time of writing is still underway, with the Lorado mill site and many of the satellite sites now having been substantially completed.

This project has been important for SRC in numerous ways. It expanded SRC's project management skills, provided experience in dealing with regulators in a more substantial way than previously (including SRC holding licences from the Canadian Nuclear Safety Commission, CNSC, for aspects of the work), represents the formation of a new line of business, and of course it has been the largest contract project ever undertaken at SRC (originally estimated at ~$24.6 million, scope expansions such as the decision to cover all of the Gunnar tailings have increased the projected value to approximately $250 million [139,140]).

Corporate, Revisited. By 2006/07 SRC had grown to about $36 million in revenue with about 329 employees; and had become the second largest provincial/territorial research organization in Canada after the Alberta Research Council (see Appendix 9.1 for a list of Canada's Provincial Research Councils). At this point (in 2007) a new strategic plan was developed that focused on continuing to increase SRC's breadth and depth in order to achieve greater economic impacts in Saskatchewan, and with additional focus on positioning to achieve significant impacts in the environment and environmental sustainability.

In 2009 SRC acquired facilities in Uranium City to support its environmental remediation activities at former uranium mines in the Athabasca Basin region (mentioned above). This also helped expand SRC's physical presence in the province from south to north (Figure 6.39).

In 2011/12 SRC's strategic plan was again revised, this time with an eight-year planning horizon. The new "Strategy 20/20" plan is still in effect at the time of writing and is still built on balanced business drivers, plus themes of broader engagement, continued diversification, and a revised Vision calling for greater international visibility and recognition [115].

One of the features of the three *New Millennium Years* strategic plans was having more balanced business drivers, especially balancing the industrial growth and support of SRC's Mandate and Mission with other needs of the province (SRC's owner). One aspect of this involved increasing the amount of direct service provided to the provincial government. In the fiscal year 1999/2000, at the close of *The Commercial Years*, SRC's contract work for the province represented 6% of the total revenue ($1.2 M of $19.2 M). By 2015/16 it had nearly quadrupled to over 22% ($15.6 M of $69.4 M), representing work in a broad range of areas.

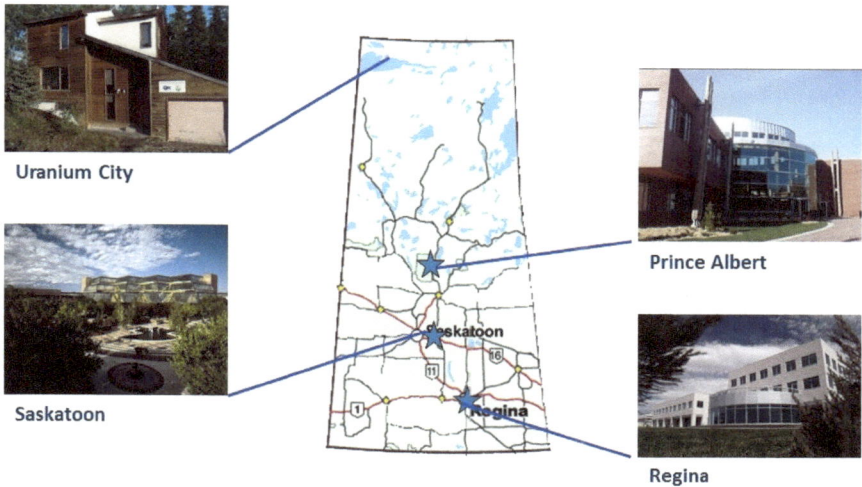

Uranium City

Saskatoon

Prince Albert

Regina

Figure 6.39. By 2009 SRC was operating multiple facilities in Saskatoon plus regional facilities in Regina, Prince Albert, and Uranium City.

With this balance in mind, another feature of the *New Millennium Years* strategic plans was growth, balanced growth. As shown in Figure 6.41, SRC's revenues in the *New Millennium Years* increased by more than three and a half times (3.5X) over those of the *Commercial Years*, and 42 times those of the *Building Years*.

Figure 6.40. In 2011/12 SRC updated its company logo in concert with its revised strategy, "Strategy 2020" [115].

Table 6.1. Illustration of some of the changes experienced by SRC during *The New Millennium Years*, 2001 – Present. See also Tables 3.1, 4.1, 5.1.

	2000	2016
Total Revenue	$19,166 k	$69,679 k
Provincial Investment	$7,947 k	$22,475 k
Contracts	$10,679 k	$46,926 k
Fixed Assets (mostly laboratory equipment and fixtures)	$6,927 k	$56,134 k
Grants (to universities)		
Grant Expense	$0 k	$0 k
Grants as % of Expenses	0%	0%
Employees	207	368
Reference	[82]	[141]

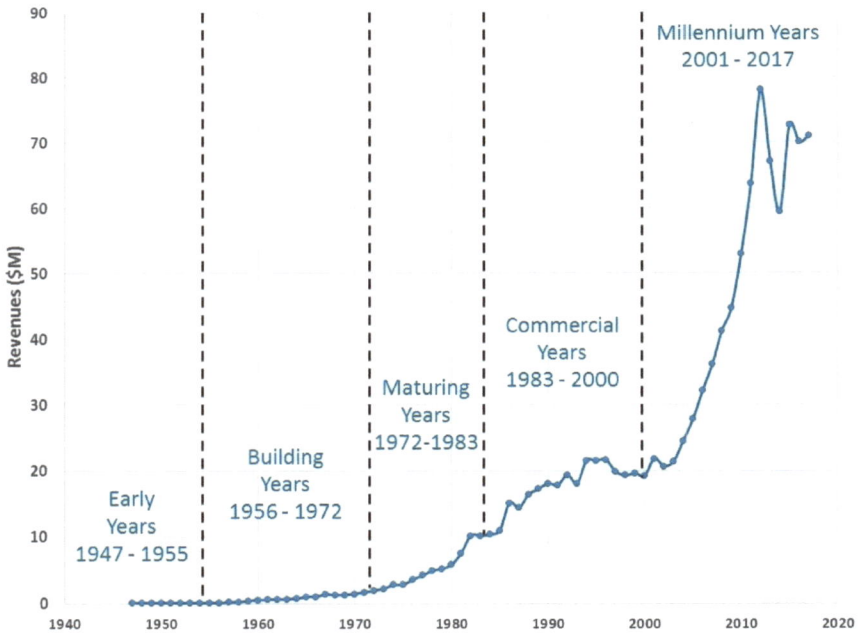

Figure 6.41. SRC total revenue growth across the eras.

7 THE IMPACTS OF SRC'S WORK

In the 20th century, Canada's research and technology organizations (RTOs) were interested in the economic impacts of their work but didn't feel that there was a way to measure them, except for the impacts of some specific projects or programs. For example:

- NRC found that it was "*difficult to estimate the costs of research and extremely difficult to estimate the returns from this investment*," however, sometimes "*a single application of scientific knowledge does lend itself, for a little while, to estimations of dollar value that are clear, simple, and convincing*" [7]. For example, in 1923, Canada's lobster industry was losing money due to discolouration of its canned lobster product. NRC's work provided a technical solution that was adopted by the industry. The actual work cost $2,000, applying the solution saved the lobster industry about $700,000 per year, a single year's savings paid back the entire cost of all of NRC's operations for its first seven years, from 1917 to 1923 [7].

- Similarly, SRC reported in 1950 that: "*It is difficult to assess the return on monies invested in research; certainly no attempt will be made to arrive at an estimate in this report ... A successful conclusion to almost any of the projects ... will offer the people of this province a potential return far beyond the amounts invested*" [27]. In SRC's 1956 Annual Report it was noted that: "*It is not possible to assess precisely the value of the applications but it is probable that they result in savings within the Province that are greater than the total expense of the Council up to the present and currently foreseen*" [33].

- This view was reinforced by Dr. John W.T. Spinks in 1958, shortly before he became President of the University of Saskatchewan: "*... the Saskatchewan Research Council has attempted to initiate and encourage research on a number of problems relating to the economic life of the Province and to the improvement of living conditions in the Province. It not expected that* *all of these will be successfully solved, but it is certain that worthwhile contributions to the solution of many of them will result ... The impact of this research ... is multiple. It advances knowledge, helps develop our provincial resources, ... helps us poor mortals reach out to the stars, and, ... helps us to do just a few of the impossible things we people in Saskatchewan have always been capable of doing"* [2].

o SRC's impacts were also recognized in 1958 by then Premier T.C. "Tommy" Douglas: *"The need for a scientific organization responsible to the Provincial Government was realized shortly after the second world war. Indications that the economy of the province was entering a phase of rapid change presenting new opportunities and problems in the fields of science and engineering prompted the Provincial Government to establish the Saskatchewan Research Council in 1947. ... the consequent benefit to the provincial economy relative to the expenses of the Council has been beyond expectation"* [1].

7.1 SRC's Early Economic Impacts.

Something every RTO has in common is an ever-increasing need to measure and demonstrate the outcomes and the impacts of their R&D work and therefore the return on investment (ROI) that their work delivers, regardless of whether the original investments came from internally or externally generated funds, from the public or private sectors, or both. NRC, SRC, and the other RTOs in Canada did take, as an article of faith, that reaching a successful conclusion to almost any of their significant research and development programs would contribute large returns on investment. As SRC's industrial work matured, some specific cases of clear economic impacts, as reported by the clients themselves, began to emerge - beginning in about 1970. The most commonly reported impacts were in product development cost savings, production cost savings, sales and export increases, job creation and layoff/bankruptcy avoidance, and new product developments and launches. Table 7.1. provides a summary of some of the economic impacts reported by SRC clients in the 1970s and 1980s. This was helpful feedback, although SRC was not usually permitted to reveal the names of the clients themselves. For example:

o In 1970, a poultry producer reported that SRC's help applying industrial engineering techniques was saving them $30,000 per year, while a tractor manufacturer reported that SRC's work was saving them $100,000 per year. These were two out of about 90 clients that year [50].

o In 1971, a client attributed $80,000 per year in cost savings due to SRC's work, while another estimated $30,000 per year in cost savings due to SRC's work with them. These were only two of about 100 clients served that year [48].

o In 1972, assistance was provided to two companies that were experiencing losses due to otherwise acceptable animal feeds being lost in waste streams. An SRC solution that enabled recovery and sales of these feeds boosting combined revenues for the two companies of nearly $100,000 per year [36].

o In 1974, three companies out of the hundreds served that year were asked to estimate the benefits, if any, of SRC's services [52]. A manufacturer of fibreglass products that received SRC's help with resistance to weathering estimated the value at $5,000 per year in cost savings for the life of the product concerned. A plastics manufacturer that received SRC's help with additives selection estimated the value at $20,000 to $30,000 cost savings. A prefabricated house building company that received SRC's help in process improvements estimated the value at $25,000 per year in cost savings.

Table 7.1. Illustration of some of economic impacts reported by SRC clients.

Year	Total Clients Served	Client Case Studies with Disclosed Impacts	Total Direct Economic Impact from the Case Studies	Reference
1970	~90	2	$130k	[50]
1971	~100	2	$110k	[48]
1972	n/a	2	$100k	[36]
1974	n/a	3	$55k	[52]
1979	200	33	$3,480k	[53]
1980	~300	19	$2,600k	[58]
1982	~300	>17	$3,215k	[63]
1983/84	n/a	33	$695k	[71]
1984/85	n/a	26	$771k	[69]

7.2. SRC's Economic Impact Assessments in the New Millennium Years.

"The voice that can credibly speak to the impact(s) of R&D is the voice of the customer."

Schramm *et al.*, 2011, [142].

In Chapter 6 it was noted that a key element of SRC's *"go-forward"* presentation to the Provincial Government in late 2001, was that SRC would commit to a value proposition amounting to the delivery of economic returns on government's investments in SRC. More specifically, the goal would be to deliver demonstrable annual economic impacts to the province that would greatly exceed the annual base funding support provided by government, and SRC's CEO projected that SRC should ultimately be able to leverage every dollar of annual base support with about another two dollars annually from industry, and with the work done with those funds achieve annual, incremental economic impacts of at least ten times the base funding amount. In the modern language of economic impacts that would mean achieving a Mandate Effectiveness Ratio (annual incremental economic impact divided by annual base funding from government) of at least 9.5 every year. As this proposition was accepted by the province, one of the next challenges was to build an appropriate economic impact assessment tool and process.

In 2002 SRC launched a project to develop an economic impact audit process that would ensure that solid data would be collected that could provide an indication of the direct economic impact of SRC's work. This wasn't done in a vacuum, but rather was built upon the somewhat similar programs that other organizations had recently developed and deployed, most notably that of John Kramers at ARC in the late 1990s [143,144]. SRC then adapted and adopted features to create a process that would suit SRC's unique role as a provider of both contract research and development and public good initiatives. The result was an R&D impact assessment tool that relies on "the voice of the customer" to provide inputs that can be aggregated and scaled-up to provide conservative estimates of R&D return on investment (ROI) [142,145]. It was first implemented in SRC's 2002/03 fiscal year [95].

The details of the economic impact audit tool and process have been described elsewhere [142,145]. Only an overview is presented here. Part of the measurement challenge is that SRC provides products and services to about two thousand clients every year and it is not practical to interview every client every year. Instead, a selection is made of past projects that have been completed for private-sector clients. The kinds of data collected included sales and other revenues, cost savings, new revenue flowing into the economy, job creation and maintenance, and increased productivity (see Figure 7.1). The bulk of the data is collected directly from clients through face-to-face interviews in which they are asked to describe if and how SRC's work had benefited their organization[56]. The interviews are usually conducted only for programs that are either mature or completed since there is generally a lag between the time an R&D project is conducted and the time at which the results have been implemented into commercial practice and business returns on investment are positive (see Figure 7.2).

The result of these interviews is a repertoire of Impact Audit Reports for each client outlining details on the work, its benefit to the client, as well as economic data. In addition to hard data, some clients are able to provide qualitative information on how well SRC's work benefited their business. Although such qualitative information cannot be easily aggregated, it does provide additional feedback on difficult to measure yet valued contributions to the clients' successes. For example, some clients have reported that without SRC's involvement their business would not have survived. Finally, a key feature of these case studies is to have a senior executive or manager from the client firm authenticate and sign-off on the reports (see Figure 7.3), making them auditable and helping to ensure the accuracy of reporting on the impacts and also on the attribution of the SRC contribution(s).

Once the data have been collected, several kinds of calculations can be performed, such as:

1. Impacts achieved in the audit year that resulted from completed SRC projects of prior years, whose results continue to produce incremental impacts. Of these, a selection of medium- to long-term 'historical' clients are interviewed. This captures the incremental jobs maintained, economic activity generated, and costs savings that such historical clients experienced in Saskatchewan, in the audit year in-hand, as a result of SRC's previous, completed work. These impact results are credited using a conservative 15 percent attribution factor. The client impact audit reports described above (and illustrated in Figure 7.3 above) provide additional context and qualitative details.

---

[56] All client information from these interviews is held confidential and is only disclosed more broadly with their permission, while aggregated and summary information are disclosed more broadly.

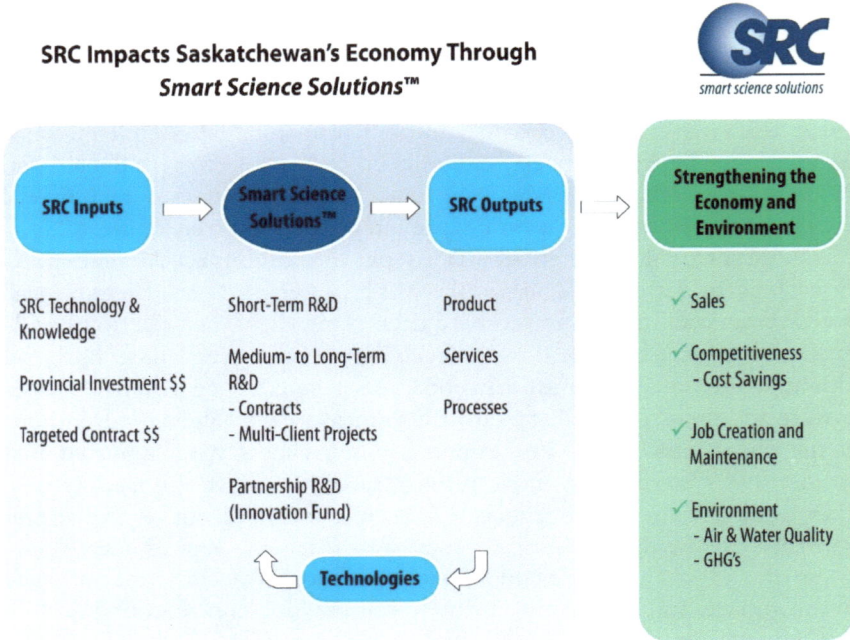

Figure 7.1. Illustration of the linkages among inputs, activities, outputs, and outcomes (impacts). From Reference [142].

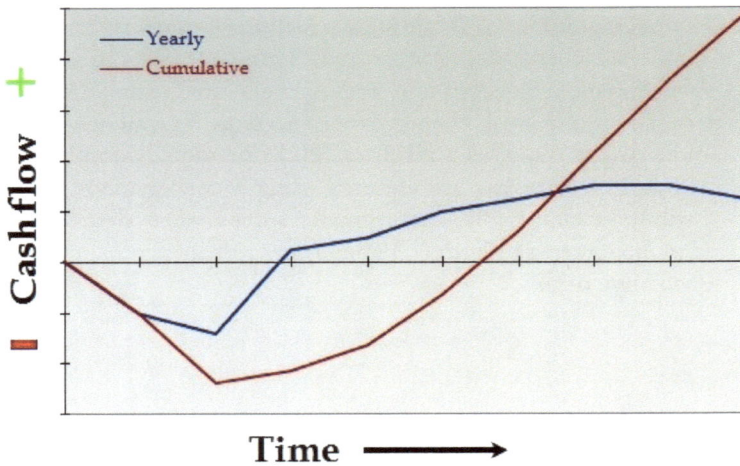

Figure 7.2. The lag-time in realizing impacts from R&D. From Reference [142].

Impact Audit Report: Hathor Exploration Limited

Technology: Uranium Analysis

Uranium is one of Saskatchewan's leading industries, and Saskatchewan is the world's largest producer and exporter of uranium. The province accounts for approximately 25 per cent of the world's uranium production and 100 per cent of Canada's production.[1] Hathor Exploration Limited is a junior uranium exploration company focused on projects in the Athabasca Basin of Northern Saskatchewan, the premier location in the world for high-grade uranium deposits. Hathor's main exploration properties are located in the eastern side of the Athabasca Basin along the same geological trend that hosts all of Saskatchewan's producing uranium mines. Hathor is currently advancing its Midwest Northeast Project, which represents the best discovery in the past 20 years by a junior uranium exploration company.

SRC provides a wide variety of services and research to the uranium and mineral exploration industry. SRC's Geoanalytical Laboratories offer comprehensive services in uranium analysis that are globally recognized and have established industry best practices. Hathor uses our Geoanalytical Laboratories exclusively for its uranium analysis.

"SRC's Geoanalytical Lab is a world leader in uranium analysis. Their quality of work and commitment to the customer are excellent."

Alistair McCready, Ph.D.
Senior Project Geologist
Hathor Exploration Limited

As a result of Hathor's uranium discovery combined with SRC's lab results, the company has significantly increased its staff from 3 to 18 people in 2008. Alistair McCready, Senior Project Geologist stated that the additional staff are mix of highly qualified geologists and junior geologists fresh from university. In addition, due to the new discovery, Hathor's share price rose from $0.44 to $2.00, and Hathor was able to raise over $55 million (net) at the increased share prices. Hathor has one of the largest land packages covering key uranium-hosting geology in a jurisdiction that has the richest uranium deposits in the world. SRC is proud of its relationship with Hathor and will continue in its commitment to set quality standards in uranium analysis.

I have reviewed this document and release its content to the Saskatchewan Research Council for the purpose of impact reporting.

Alistair McCready, Senior Project Geologist
Hathor Exploration Limited

---

[1] Ministry of Energy and Resources, Government of Saskatchewan, http://www.er.gov.sk.ca/Default.aspx?DN=3564,3541,3538,3385,2936,Documents, 22 April 2009.

Figure 7.3. Example of an SRC Economic Impact Audit Report. Courtesy of Hathor Exploration Ltd., 2009.

2. Impacts achieved in the audit year that resulted from active SRC projects that were either completed in the audit year or are on-going, and whose results produced incremental economic impacts. Again, a selection of (usually) multi-year, science-based technology projects that were active during the audit year is evaluated. Where multi-client, consortium-style projects are involved, each client is asked to provide their assessment of the benefits that have accrued to their particular company. To maintain a credible cross-section of projects, the basket of projects changes from year to year with the addition of new projects and the removal of others. This basket of projects provides the backbone for SRC's economic impact analysis because they are used to extrapolate from these results in order to estimate the impacts from the full suite of such projects conducted in any given year.

    From these results, a ratio is calculated that reflects SRC's total applied research, development, design, testing, demonstration and technology transfer activity (for the audit year; in dollars) divided by the aggregate level of activity from the specific projects chosen for the audit (also in dollars): (Total R&D portfolio costs) / (Costs for the audited projects). This ratio is multiplied by the total amount of direct client impact for the audited projects, in terms of each of jobs, cost savings, and new revenues flowing into the Saskatchewan economy. The result represents SRC's overall involvement in the Saskatchewan economy during the audit year, but not SRC's direct economic impact. The actual direct economic impact is calculated by applying an attribution factor.

    The need for an attribution factor comes from the fact that most of SRC's projects involve one or more partners and also because the ultimate impacts invariably involve contributions from the clients themselves. Attribution is very difficult to quantify, however, the importance of SRC's contributions to a client's success is one of the questions asked during the audit interviews, and based on direct client assessment a conservative overall estimate can be obtained. SRC has been using a very conservative 15 percent attribution factor. Once again, the client impact audit reports described above provide additional context and qualitative details.

3. SRC's direct economic activities in the audit year, in terms of SRC jobs and SRC revenues. The impact results in this category are assessed using a 100 percent attribution factor.

The impacts from the above three categories, adjusted by their respective attribution factors, are then summed to yield an overall economic impact value for the audit year. Since its inception in 2002/03, the economic impact audit process has been institutionalized so that, in addition to providing each year's update, SRC is also able to follow trends such as are shown in Table 7.2 and Figure 7.4. These show that between 2003 and 2016 SRC achieved over $7.1 billion in economic and employment impacts in Saskatchewan.

SRC's impact on Saskatchewan's economy in fiscal year 2015-16 was more than $484 million. SRC also assisted in the creation or maintenance of more than 4,826 Saskatchewan jobs. The value of these jobs in terms of economic impact that year was nearly $329 million.

Table 7.2. SRC Impacts from 2003-04 to 2015-16.

	Jobs	$ Impact (M)	Mandate Effectiveness
2003-04	1,181	$208	26
2004-05	1,907	$268	34
2005-06	2,450	$343	40
2006-07	2,968	$400	46
2007-08	1,940	$324	36
2008-09	738	$447	37
2009-10	939	$542	36
2010-11	1,206	$527	32
2011-12	1,894	$656	36
2012-13	2,701	$559	29
2013-14	1,927	$413	21
2014-15	832	$519	26
2015-16	4,826	$484	22

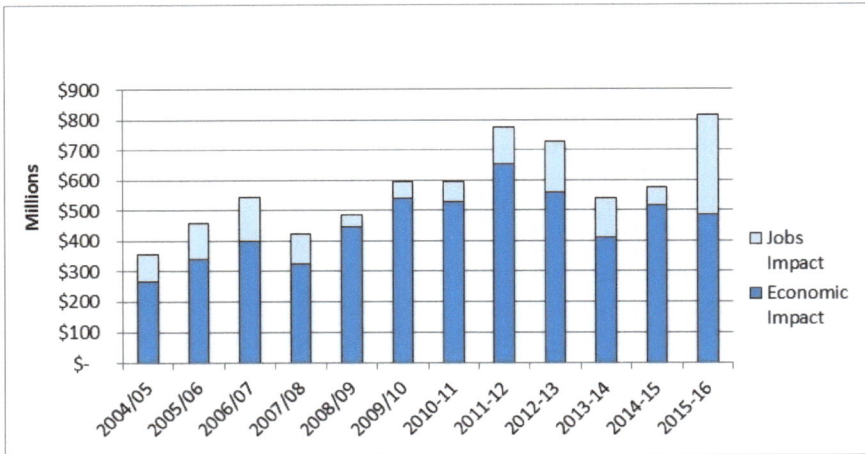

Figure 7.4. SRC's Annual Economic Impacts in Saskatchewan.

SRC's Mandate Effectiveness Ratio (sometimes referred to in the literature as mission effectiveness) is the leverage that SRC creates in the province with the government's base annual investment[57] and is equal to SRC's annual incremental economic impact divided by the government's annual 'base funding' investment in SRC.

The data in Table 7.2 show that between 2003 and 2016, SRC's mandate effectiveness ranged from a low of 21 to high of 46, with an average of 32.4. This means that for every $1.00 that the government invested in SRC, its work contributed a return on investment of at least $32 of growth in the Saskatchewan economy. This is more than double the level of 9.5 proposed to government in 2001 and, if the impacts of annual job creation and saving are added in as well, the total has been increasing over the years.

SRC's Annual Direct Returns to Saskatchewan's General Revenue Fund. In 2015, an estimate was made of another kind of SRC impact: the annual direct economic return to Saskatchewan's General Revenue Fund (GRF) as a result of SRC's work with Saskatchewan-based and Saskatchewan-operating clients [146]. This is related to, but distinct from SRC's annual impacts contributing to overall economic (GDP) activity and job creation/savings activities as described earlier in this section. This assessment considered three types of annual GRF impacts that derive from

[57] The base annual investment is the direct, non-targeted 'base funding' provided by the province, as distinct from government revenues arising from specific contract R&D projects.

SRC's work and was calculated using average corporate and personal tax data for the 2011 and 2012 tax years, and conservative SRC impacts data for 2014/15:

1. Direct incremental business activity impact, representing the corporate income taxes paid to the province by SRC clients as a result of their incremental business revenues attributable to SRC's work ($ 4.1 million),

2. Direct jobs impact, representing the personal income taxes paid to the province by SRC employees themselves and the employees of SRC clients, where their jobs have been created and/or saved as a result of business activities (revenue growth and/or efficiency and/or productivity increases) attributable to SRC's work ($ 4.6 million),

3. Royalty revenue impact, representing the incremental resource royalties paid to the province by SRC clients based on their natural resource extraction activities in oil, gas, potash, and uranium to the extent attributable to SRC's work in aid of such clients ($14.9 million).

The three categories, taken together, provide an estimate of annual direct economic returns to Saskatchewan's GRF as a result of SRC's work with Saskatchewan-based and Saskatchewan-operating clients as $23.6 million in 2014/15. The government's investment in SRC from the GRF that same year was $20.3 million, so this estimate suggests that incremental taxes and royalties paid to the government as a result of incremental and profitable business activities enabled by SRC's work more than pays back the annual base investment in SRC – and this is in addition to the direct economic impacts described earlier.

7.3. Selected Economic Impact Case Studies.

The Implementation of Horizontal Wells[58]. In the late 1980s, the concept of drilling horizontal[59], rather than vertical, wells in order to improve access to productive regions in petroleum reservoirs was gaining interest on the part of Canadian oil producers.

Horizontal wells had been used occasionally in the United States since the 1930s, but the necessary drilling technology really only became broadly practical in the early- to mid-1980s [147]. Sceptre Resources (now Canadian Natural Resources ltd., CNRL) approached SRC for assistance in designing the best well arrangement, predicting well performance, and determining the best operating strategies.

SRC determined the best way forward was using scaled physical model techniques. Numerical models using computers were not sophisticated enough at the time to model steam injection into horizontal wells, so Brian Kristoff and others at SRC adapted an existing scaled-physical-model laboratory that had originally been built for vertical well scenarios. A three-dimensional scaled-physical-model specific to the Tangleflags Field (in southern Saskatchewan) was created to model a pattern containing a horizontal producing well with multiple vertical steam injection wells. A program of studies using the new physical model was used to predict well performance, determine the best operating strategies, and perform financial analysis for a possible field pilot test.

Based on positive results Sceptre drilled Saskatchewan's first horizontal well in 1987 at the Tangleflags North field [148]. The pilot was successful and this well operated for many years under steam injection, and the predictions provided by SRC's 3D scaled physical model turned out to be within a few percent of the actual field results. This single horizontal well at peak production accounted for one percent of the Saskatchewan's total production from 15,000 wells. By the end of its operating life (in 2013) this one well had alone produced an estimated 3.46 million barrels of oil [149]:

"*[F]or the Tangleflags North Lloydminster heavy oil pool ... Saskatchewan Research Council ... was engaged to provide services in support of designing a steam flood field test to demonstrate the potential for oil recovery improvement ... [with] horizontal wells ... Highly encouraging model results led to the drilling of Saskatchewan's first horizontal well in October 1987 followed by initiation of steam*

---

[58] Many of the details in this summary were kindly provided by past SRC employees Ernie Pappas and Brian Kristoff. Their help is gratefully acknowledged.

[59] Horizontal wells need not be horizontal. In general, the term "horizontal wells" refers to wells drilled at any angle other than vertical. Drilling horizontal wells is also referred-to as "directional drilling."

injection in 1988. SRC then provided ongoing support for operating strategy optimization ... To date, this first horizontal well has produced 3.4 million barrels of oil, a testament to what is possible when strong applied research support aligns with a good reservoir, appropriate technology and a strong corporate commitment to make the project succeed."

Paul Jespersen, Manager, Enhanced Recovery,
Sceptre Resources [150].

SRC now had a field-proven predictive tool, but there remained many questions surrounding the mechanics of horizontal well production and the application of horizontal wells to different reservoir types. To address these questions and provide appropriate engineering design methods, SRC established a multi-client R&D program in 1990 that continued through to 2000. This program was very successful, and its results were quickly adopted by the petroleum industry and applied to field operations in virtually all reservoir types in the province. This had a profound effect on Saskatchewan's oil production (see Figure 7.5). Whereas three years after the first horizontal well about 5% of Saskatchewan's oil production was attributed to horizontal wells, by 1995 35% of Saskatchewan's oil was being produced by horizontal wells. By 2000 Saskatchewan was able to celebrate a doubling of oil production during the previous decade. In 2013, nearly 49 million barrels, about 26% of Saskatchewan's total oil production was by horizontal wells [123].

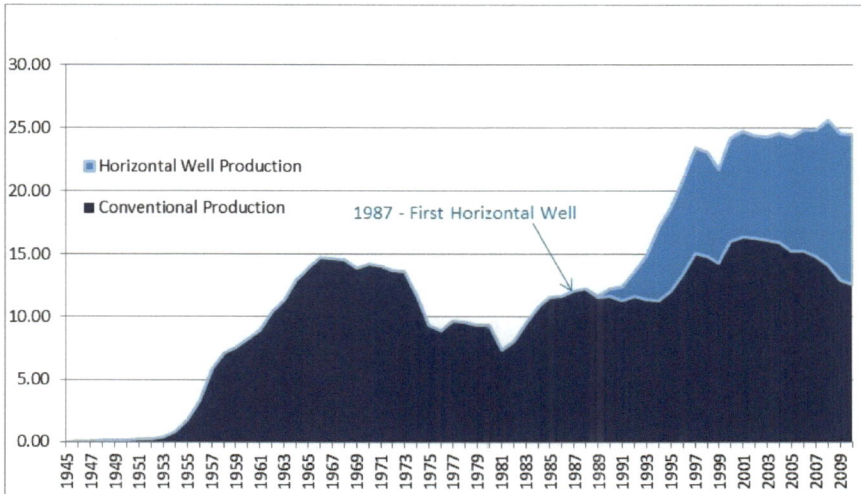

Figure 7.5. Saskatchewan oil production showing the impact of the introduction of horizontal wells (million m³/year).

The Weyburn CO_2 EOR Process. In 1986 SRC launched a multi-client project to determine if and under what conditions carbon dioxide (CO_2) would be the most appropriate enhanced oil recovery (EOR) process for a waterflooded light oil reservoir [64]. By 1997, through a combination of laboratory and numerical simulation studies, SRC found the answers to these questions and determined how to make CO_2 EOR both technically and economically feasible for some reservoirs. Based on SRC's results, EnCana Resources (formerly PanCanadian Petroleum) decided to pursue the development of a $1.1 billion EOR project in the Weyburn oil field in southern Saskatchewan [87]. At that time (1997) the Weyburn field was nearing the end of its production life having exhausted primary production, conventional waterflooding, and also the implementation of vertical "infill" well recovery and production methods. SRC helped EnCana to adapt its models and data to the Weyburn field situation, as a result of which EnCana estimated that launching a CO_2 EOR process in this field could produce an additional 122 million additional barrels of oil from the reservoir, thus prolonging economic production by about 25 years (Figure 7.6). EnCana conducted a successful field demonstration project in 2000 and had deployed the technology in about a third of the field by 2003 and full-field by about 2008. SRC continued to assist EnCanada throughout the evolution of the field implementation. Since 2000, Weyburn oil production has increased at a steady rate through 2014.

EnCana has credited SRC with being a critical enabler of the Weyburn CO_2 EOR process that they implemented in the field, saying:

"Without the work SRC did we could not have gone to the field. We needed the fundamental laboratory data to justify the capital expenditure of a commercial stage oil field operation" (Andrew Graham, EnCana Resources [151]).

Taking an average oil price of US$50/barrel for illustration, the incremental oil recovered from the CO_2 EOR phase by the time the field is abandoned in about 2030, the direct economic impact will have been about US$6 billion. Another major impact from this is the millions of tonnes of CO_2 injected, and which will be left in the reservoir at abandonment, which otherwise would have been released to the atmosphere. It has been estimated that the ultimate amount of CO_2 that can be left stored in this reservoir is 23 million tonnes [152].

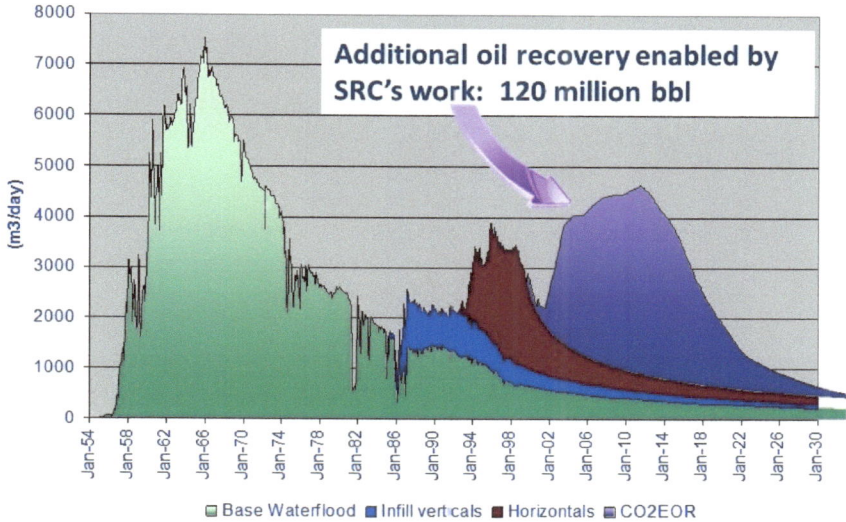

Figure 7.6. Historical and projected oil production from EnCana's Weyburn operations, showing the extensions of the reservoir's productive life due to the introduction of infill vertical wells, then horizontal wells, then CO$_2$ EOR. Based on data provided by EnCana.

7.4. Environmental, Sustainability, and/or Social Impacts.

In addition to direct economic and job-creation benefits, many of the projects SRC works on achieve positive non-financial impacts like environmental, sustainability, and other social benefits (including human and animal health and safety). Such impacts are extremely difficult to quantify but are still very important. For SRC they reflect achievement of the portion of SRC's Mission that relates to the quality of life and a 'secure environment' and are therefore also very important to the company and its stakeholders. Examples of such quality of life contributions abound:

o A non-economic impact example is the province-wide groundwater observation well network that SRC established in 1964 to provide a means of assessing and monitoring the province's huge groundwater resource. The result was a comprehensive and unique historical database of groundwater activities. The data is used by a wide range of stakeholders, including climatologists, farmers, hydrogeologists, natural gas and oil well drillers, various levels of government, and city planners. The program helps stakeholders ensure they have a sustainable fresh water supply while providing a means to minimize the risk of groundwater contamination, thereby helping to ensure a secure water supply.

o Another non-economic impact is SRC's contribution to the reduction of greenhouse gases through such work as the development and deployment of CO_2 injection technology into petroleum-bearing reservoirs for enhanced oil recovery. SRC played a leading role in the development of this technology which is now being used in the Weyburn (Saskatchewan) oilfield and which is now part of an international CO_2 storage and monitoring project. Already, many millions of tonnes of CO_2 have been injected at this site that would have otherwise been released to the atmosphere. Related projects are ensuring that CO_2 becomes sequestered in soils and plants through improved agricultural and forestry practices.

SRC began to attempt to account for at least some of these kinds of impacts of its work in 2006. Some of the methods used are described in references [142,145]. In some areas, only examples can be provided due to the difficulty of quantifying the benefits, such as in improved air quality, water resources, health, and safety. In order to demonstrate some activities producing quantifiable impacts, a basket of projects is examined each year to identify the amount of GHG reductions and/or energy savings from electricity and natural gas, CO_2 sequestration and direct emissions reductions. The common thread for these various impacts is that they can all be converted to quantifiable reductions in GHG emissions.

Table 7.3. Some of SRC's non-financial impacts from 2006-07 to 2015-16.

	R&D Projects Contributing Socio-Environmental Impacts	Energy Savings (kWh/yr)	GHG Emission Reduction (tonne/yr)
2006-07	> $12 million	> 56 million	> 24,000
2007-08	> $14 million	> 58 million	> 25,600
2008-09	> $13 million	> 29 million	> 10,000
2009-10	> $26 million	> 27 million	> 10,000
2010-11	> $37 million	> 24 million	> 9,000
2011-12	> $53 million	> 43 million	> 22,000
2012-13	> $36 million	> 44 million	> 22,000
2013-14	> $17 million	> 40 million	> 21,300
2014-15	> $27 million	> 40 million	> 21,000
2015-16	> $22 million	> 40 million	> 21,000
Sum	> $257 million	> 411 million	> 166,000

7.5. Selected Environmental, Sustainability, and/or Social Impact Case Studies.

Energy Efficient Housing and R-2000. Residential energy use is a significant fraction of energy use in many countries, including Canada for which it represents nearly 20% of total energy use. SRC's Building Performance Business Unit, led for many years by Dr. Rob Dumont (Figure 7.7), developed, adopted, and adapted, and then demonstrated numerous energy efficiency technologies in a series of commercial demonstration homes in Saskatchewan over a period of three decades.

The 1977 *"Saskatchewan Conservation House"* in Saskatoon (see Figure 4.25 above), demonstrated efficient methods for energy conservation and heating, including airtightness, "super" insulation, solar heating, and a heat recovery system. With this demonstration house, SRC was able to show that practical insulation and conservation practises could reduce home energy requirements by up to 90% [56]. SRC was later co-recipient of a Conservation Housing Design Council Award of Merit, in 1982 [63]. SRC also established six retrofit demonstration houses in Yorkton, Moose Jaw, Swift Current, Weyburn, Prince Albert, and Melville. The retrofit houses demonstrated that 40% energy savings could be easily achieved [54]. The learnings from these and other conservation projects were freely disseminated to the home building industry and others. The key learnings from this project formed the basis for the Canadian R-2000 program, which was launched in 1982.

Figure 7.7. SRC mechanical engineer and building energy efficiency expert Dr. Rob Dumont in 2007. Saskatchewan Research Council photo.

Between 1980 and 1982, a group of organizations including SRC built a series of "super" insulated houses in Saskatoon, including the so-called "*Chainsaw Retrofit House*[60]" in 1982. The learnings from these demonstrations, plus those from the 1977 "*Saskatchewan Conservation House*" discussed above, led to the design and launch of the Canadian R-2000 residential building standard program in 1982 [153,154,155].

---

[60] Referring to the "chainsaw retrofit" methodology, not the use of a chainsaw.

According to Rob Dumont: *"The [R-2000] program had a strong emphasis on performance, as opposed to prescriptive standards for energy use for space heating. … The first group of monitored R- 2000 houses consumed about 57% less energy than conventional houses"* [156]. This building standard is still in place in Canada [157].

The 1993 *"Saskatchewan Advanced House"* in Saskatoon (see Figure 5.11 above), demonstrated even more efficient methods. SRC was one of the contributing partners in this demonstration, which achieved a 50% reduction in water consumption, 75% reduction in space heating, 75% reduction in space cooling, and a 50% reduction in other electricity use [77,81]. At its grand opening, it was billed as *"One of the most Energy-Efficient, Environmentally-Friendly Homes in the World"* [81].

In 1999, with the *"Dumont House,"* Dr. Dumont won a national award for designing, building, and demonstrating a leading-edge energy efficient house consuming only about one-third the energy of R-2000 rated houses [82]. In this case, it was his family's personal residence that was used as the test/demonstration site of the same technologies with which SRC was helping local industry and communities (Figure 7.8).

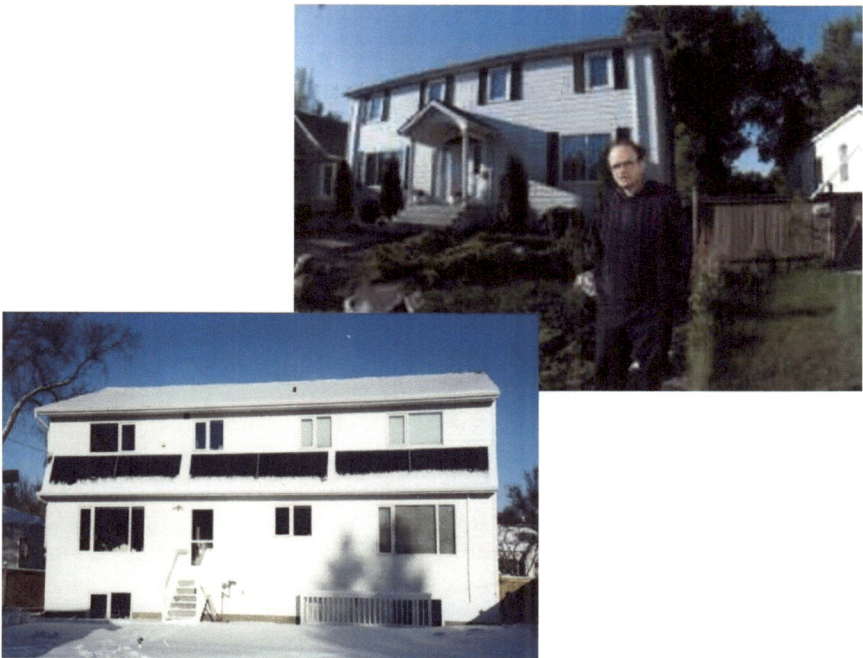

Figure 7.8. Dr. Rob Dumont in front of his award-winning Saskatoon home in 1999. Saskatchewan Research Council photos.

SRC and OEC led the design and construction of the "*Factor 9 Home: A New Prairie Approach,*" in Regina (see Figure 6.26, above). This demonstration home uses 90% less energy and 50% less water than typical homes built with 1980s technology, with an incremental cost over conventional technology of only 10% or less [158,159]. It was officially opened by Premier Lorne Calvert and SRC in 2007.

The "*Agency Chiefs Centennial Home*" at Christopher Lake, near Prince Albert (see Figure 6.27, above) was developed by SRC and OEC in conjunction with the Agency Chiefs Tribal Council (through AC Realty) with the specific needs of northern and Aboriginal communities in mind. The home, which opened in 2006/07, achieved a 35% reduction in energy use but about the same cost compared with conventional construction [97,158,160].

Lorado Mill-Site Remediation. Project CLEANS was introduced in the previous chapter. The first of the large orphaned, abandoned Cold War-era uranium sites to be remediated by SRC was the Lorado mill-site. After closure in 1961, the Lorado site was abandoned with little to no decommissioning, leaving behind hundreds of thousands of acidic mill tailings covering about 14 ha, and which had overflowed into nearby Nero Lake [136]. The principal risks to humans, wildlife, and aquatic life arose from the exposed tailings and contaminant flows from Nero Lake to nearby Beaverlodge Lake.

An engineered soil cover system was designed and placed over the entire exposed tailings area by SRC (comprising about 250,500 m^3 of sand and about 93,400 m^3 of till). This ensures protection against gamma radiation and radon exposure, prevents the formation of efflorescent salts on the surface, reduces tailings acid generation in Nero Lake, and blocks uptake of tailings pore fluid into vegetation. The cover system was designed to ensure that it is free draining and contoured to shed water towards one of two engineered drainage ditches, or directly to Nero Lake (Figure 7.9). The cover system is currently being stabilized and revegetated by SRC.

Nero Lake is about 2.2 km long by about 1 km wide, with a total water volume of about 11 million m^3. About 167,000 m^3 of tailings and acidic waste overflowed into Nero Lake and as much as 40% of the bottom of Nero Lake had been covered with tailings. This contamination was enough to cause the lake pH to drop to 4.0 (i.e., making it acidic), reduce the alkalinity and bicarbonate concentrations, and increase the dissolved heavy metal concentrations - such that the lake was no longer able to sustain a fish habitat. A treatment process for the lake was developed, pilot tested, and implemented by SRC in 2013-2014, which involved treating the lake with about 400 tonnes of lime (CaO) to bring the pH of the lake up to about 7.5 (near-neutral).

Figure 7.9. Aerial view of the Lorado tailings covering operation underway in 2014 [136].

The above measures were designed and implemented to alleviate concerns that the Lorado mill site was affecting human health and the environment. SRC's Lorado remediation plan was based on proven scientific methods, risk assessment, and health, safety and environmental best practices. Revegetation is underway as this book was being written, and monitoring will continue in future years to ensure public and environmental safety.

Gunnar Mine, Mill, and Town-Site Remediation. The largest and most complex of the 37 orphaned, abandoned Cold War-era uranium mine and mill sites in Northern Saskatchewan was the Gunnar Mine, Mill, and Town-Site. Noteworthy for its size and technical issues, this cleanup project has also achieved huge impacts in the areas of public safety, regional employment opportunities, and environmental protection.

The public safety issues arose from the many hazards present on-site and the frequent use of the site by local residents and other visitors for sight-seeing, picnicking, fishing, hunting, and transportation (via the nearby Gunnar airstrip). The principal hazards included: structural hazards (from the many standing but deteriorating and unstable structures), radiation hazards distributed across the entire site (the main radiological hazards being from uranium-238, thorium-234, thorium-230, radon-222, and radium-226), asbestos hazards (virtually all of the mine, mill, and town-site buildings had asbestos-containing roof coverings, exterior coverings, insulation, interior coverings, and litter), and chemical hazards including large quantities of spent and unspent process chemicals, large piles of elemental sulphur, hundreds of PCB-containing lighting ballasts, and some 8,000 barrels in various locations containing unknown materials ... [138].

The regional employment opportunities arose out of SRC's work to maximize local benefits during the many years of this remediation project. For example, during the Gunnar demolition phase (2010 – 2011), in which all of the standing structures of the town-site, mine, and mill were demolished, SRC organized training for about 130 Athabasca Basin residents to ensure that all local residents that desired employment on the project were qualified for it. In 2011, half of the demolition workforce at Gunnar was made up of Athabasca Basin residents [138]. In subsequent phases of the Gunnar project, such as the remediation of the tailings phase (2016 – 2020) SRC will again be working to maximize training and employment opportunities for the Athabasca Basin communities.

Figure 7.10. The Gunnar mine and some of the mill buildings before demolition. Photo courtesy of Woodland Aerial Photography.

The environmental issues included altered landscape, altered vegetation, altered groundwater (introducing chemical and radiological hazards), altered water bodies and their sediments (introducing chemical and radiological hazards), and air pollution (from radioactive dust). The three large Gunnar tailings areas are currently being remediated (2016 – 2020), and remediation plans are being formulated for the demolition debris, the flooded mine pit, and the remaining waste rock. [138].

7.6. Corporate Social Responsibility.

All of SRC's work is aimed at benefitting the people and communities of Saskatchewan, mostly by helping the people, communities, and businesses of the province to strengthen the economy, with quality jobs, in a sustainable manner with respect to environmental health, safety, and security, and also with respect to the sustainable development of resources. In order to accomplish this SRC strives to balance its economic, environmental, and social objectives.

SRC's first corporate social responsibility (CSR) strategy was developed and pilot tested in 2007/08, leading to a publicly referenced CSR Strategy being refined and launched in 2008/09 [106]. SRC's CSR Vision has been to become recognized as a strong business, known for integrity, safety, employee and client relations, environmental leadership and community engagement. In order to accomplish this, SRC has strived to achieve best practices and continuous improvement in six key, interrelated, areas (Figure 7.11): (1) Safety and Health, (2) Environment, (3) Governance and Business Practices, (4) Employee Engagement, (5) Partner, Supplier, and Client Engagement, and (6) Community Engagement.

Figure 7.11. Six key, interrelated areas in Corporate Social Responsibility (CSR).

Among the various sub-strategies and tactics developed and deployed in 2009 and subsequent years, a common theme was to report to both employees and the public on SRC's CSR initiatives and their outcomes and impacts, such as in SRC's Annual Reports. In 2010 SRC adopted the internationally accepted Global Reporting Initiative™ (GRI) standards for corporate social responsibility as a way to stay accountable and on-track while continuing to advance and grow the organization, including annual public reporting [161-165]. Other advances have been an increasing use of social media to communicate with a broad range of stakeholders (Figure 7.12) and reaching out to students (see Figure 7.13 and examples provided below).

Figure 7.12. Vice President Wanda Nyirfa explains SRC's approach to Corporate Social Responsibility in a video for social media in 2012 [168].

Although there are many facets to CSR, as noted above, many people think of community activities when they think of CSR. Under the CSR strategy in this area, through an employee-driven CSR committee, SRC began to coordinate both organizational activities and also engagement with and support for many of the volunteer activities that SRC employees were already doing on their own. Examples of such activities have included SRC-sponsored scholarships, participating in local community-based charitable activities and drives (like shoreline cleanups, recycling drives, and Saskatoon's annual Dragon Boat Races), and supporting the Saskatchewan food banks and the like [82].

> *"No one else compares to SRC in terms of volunteer hours ... You're building a model for how other organizations can engage. You are the model."*
>
> Laurie O'Connor, Executive Director,
> Saskatoon Food Bank & Learning Centre, 2014 [164].

Other, more corporately driven activities have included hybrids of business and community support such as SRC's partnerships and support for organizations like the Conservation Learning Centre (CLC), and for Saskatchewan's Aboriginal and Northern communities. Still other initiatives have involved SRC, province-wide communities, educational activities and skills development, such as the following examples.

Figure 7.13. Illustration of a Take Our Kids to Work Day video prepared for social media in 2016 [169].

Canada's Biggest Science Experiment. SRC has often developed programs to expose students, usually at the Grade 5 and Grade 6 levels, to hands-on science experiments. For example, in 2000/01 through 2002/03, SRC held open-house "Innovators in the Schools" events in its Regina and Saskatoon facilities, which typically involved about 200 students per event [95].

In order to engage a larger number of students, and from communities across the province, a larger event dubbed *"Canada's Biggest Science Experiment"* was developed. In 2003, under this program, SRC employees guided over 1,700 "honourary scientist" students from across Saskatchewan in an experiment involving the testing and comparison of the stability of foams made from household "soaps" (detergents) [166]. In 2005, SRC employees guided over 3,000 Grade 5 and 6 "honourary scientist" students from across Saskatchewan in an experiment involving water testing using samples of water from their own communities (Figure 7.14) [105,167].

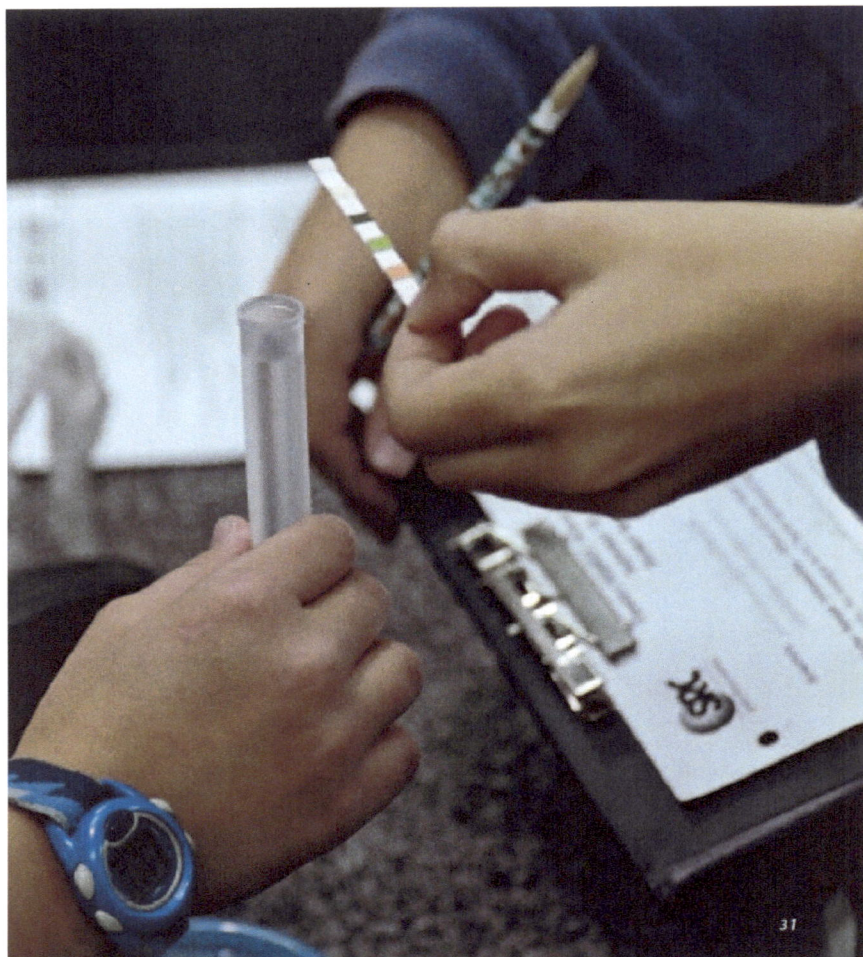

Figure 7.14. Conducting an experiment on a water sample during SRC's Canada's Biggest Science Experiment in 2005.

These programs provided thousands of students with hands-on opportunities to conduct real science experiments, to see how science affects their daily lives, and to experience some of the excitement that a career in science can provide. Another goal was to encourage student interest in science, technology, engineering, and math.

SRC - Saskatchewan Polytechnic - Northlands College Chemical Technician Program. In 2006/07, SRC partnered with the Saskatchewan Institute of Applied Science and Technology (SIAST, now called Saskatchewan Polytechnic), and Northlands Regional College to develop and launch a new program aimed at helping to prepare Aboriginal students for careers in science [97]. Under this program, students from northern Saskatchewan can pursue training in the Chemical Laboratory Technician program at Northlands College in La Ronge, Saskatchewan. Following the first year of studies, SRC offers the students summer employment in its analytical laboratories in Saskatoon. The students can then complete their final year in Saskatoon at Saskatchewan Polytechnic, after which they can receive their Chemical Technology diploma certification. The 2-year program includes a combination of classroom and laboratory instruction, fieldwork, and on-the-job training at SRC.

Aboriginal Mentorship Program (AMP). A mentorship program for Aboriginal post-secondary science, technology, engineering and math (STEM) students that includes coaching and mentoring throughout the school year, plus relevant summer employment, to help set students up for success. The aim is to provide meaningful summer employment and opportunities to help First Nations, Inuit and Métis students to develop skills and gain experience that will guide them in their studies and future careers. AMP is led by SRC in collaboration with the University of Saskatchewan, Gabriel Dumont Institute, and Saskatoon Tribal Council [141].

Some examples of public milestones by SRC reached include:

o **2010/11**. Saskatchewan Employer of Excellence Award, Large Employer category, Sask. Association of Rehabilitation Centres; SWWA Supplier of the Year Award, Saskatchewan Water and Wastewater Association Annual Conference; Award of Excellence in Communication Management, Saskatchewan Awards for Communications Excellence (ACE); Saskatchewan's *Top 100* Companies Award, Saskatchewan Business Top 100.

o **2011/12**. Recognized by the American Society for Quality and KAIZEN Guru Masaaki Imai for promoting Lean philosophies for economic development in Saskatchewan; *Saskatoon Shines!* Tourism Leadership Award, City of Saskatoon and Tourism Saskatoon; Saskatchewan's *Top 100* Companies Award, Saskatchewan Business Top 100.

o **2012/13**. *Silver Leaf Award for Excellence*, International Association of Business Communications (IABC); two Gold Level MarCom

2012 awards (from the International Competition for Marketing and Communication Professionals); three AVA Digital Awards; Certificate of Recognition, Lung Association of Saskatchewan (for radon testing services and awareness campaign); Environmental Excellence Award, Association of Professional Engineers and Geoscientists of Saskatchewan (APEGS); Saskatchewan's *Top 100* Companies Award, Saskatchewan Business Top 100.

o **2013/14**. *Future 40 Responsible Leaders* list, Corporate Knights; Leadership in Community Service *Communitas Award* from the Association of Marketing and Communication Professionals; 2013 WebAwards *Standard of Excellence Award* (science category), Web Marketing Association.

o **2014/15**. *Future 40 Responsible Leaders* list, Corporate Knights; *Canada's Outstanding Employers Award,* The Learning Partnership.

o **2015/16**. *Future 40 Responsible Leaders* list, Corporate Knights; *Work-Life 2016 Seal of Distinction,* WorldatWork; *Canada's Outstanding Employers Award,* The Learning Partnership; Saskatchewan's *Top 100* Companies Award, Saskatchewan Business Top 100.

o **2016/17**. *Mission: Zero Award* (first-place, large employer category), Safe Saskatchewan; certified *Great Place to Work* in Canada, Great Place to Work organization; *Canada's Top 100 Employers Award,* Canada's Top 100 Employers organization; *Canada's Top Employers for Young People Award,* Canada's Top 100 Employers organization; Work-Life 2017 Seal of Distinction, WorldatWork; *Canada's Outstanding Employers Award,* The Learning Partnership; Saskatchewan's *Top 100* Companies Award, Saskatchewan Business Top 100.

7.7. The "Bottom Lines."

As discussed in the earlier sections of this chapter, SRC did not start to regularly and systematically assess its impacts in Saskatchewan (and beyond) until 2003. Some quantitative information on SRC impacts for prior years is available, however, and can be used to make a very conservative estimate of SRC's impacts in Saskatchewan over its first 70 years.

For the years 1947 through 2002/03 SRC's direct activities, as measured by cumulative total revenue, was about $437.6 million. Some direct economic impacts were reported for some of these years, as discussed in Section 7.1. Although these do not represent all impact nor all years in this range they do at least provide some direct impacts as reported by SRC's clients. For the examples cited in Section 7.1, the cumulative value is about $11.1 million. Therefore, an extremely conservative estimate for SRC's economic impacts in the period 1947 through 2003 would be not less than

$448.8 million. As discussed in Section 7.2, for the years 2003/04 through 2015/16 SRC's direct economic impacts plus jobs impacts was about $7.1 billion (about $7,154.3 million). A conservative estimate for SRC's total economic impacts for it entire history to date would thus be at least $7,603.1 million.

The total provincial investment in SRC from its inception in 1947 to the end of the fiscal year 2015/16 was about $362.3 million. The overall mandate effectiveness or economic return on the provincial investment for SRC's work from 1947 through 2015/16 is at least $7,603.1 M divided by $362.3 M, which comes to 21.

SRC's lifetime return on investment (ROI) in terms of enabling economic growth in Saskatchewan is over 21.

Some accumulation of SRC's socio-environmental impacts can be made although, as discussed in Section7.4 above, these have only been systematically evaluated in recent years. Nevertheless, for the ten-year period for which data is available: 2006/07 through 2015/16, SRC conducted over $257 million worth of R&D projects contributing socio-environmental impacts that include:

o   Over 411 million kWh in cumulative energy savings, enough savings to power over 12,000 Saskatchewan homes for a year, and
o   Over 166,000 tonnes in cumulative GHG emissions reduction, enough to offset ~24% of Saskatchewan cars for a year[61].

The above economic and socio-environmental impacts refer only to SRC's impacts within Saskatchewan. Only a little attention has been paid to SRC's impacts in Alberta, the rest of Canada, and around the world. These broader impacts exist, as reported by SRC's clients, and range from enabling the development of pipeline slurry hydrotransport for Alberta's mineable oil sands operations, to developing enhanced oil recovery process modifications for oil producers in the United States, to assisting with international diamond exploration. For the most part, however, aggregate quantitative estimates of such ex-Saskatchewan impacts remain to be assessed.

---

[61] Based on Saskatchewan home energy use and vehicle emissions data from Statistics Canada for 2007; mean number of vehicles per Saskatchewan household data from Natural Resources Canada for 2008; and number of vehicles weighing under <4,500kg from Climate Change Saskatchewan.

LAURIER L. SCHRAMM

8 RETROSPECTIVE

"The need for a scientific organization responsible to the Provincial Government was realized shortly after the second world war. Indications that the economy of the province was entering a phase of rapid change presenting new opportunities and problems in the fields of science and engineering prompted the Provincial Government to establish the Saskatchewan Research Council in 1947. ... the consequent benefit to the provincial economy relative to the expenses of the Council has been beyond expectation."

Premier T.C. "Tommy" Douglas, 1958 [1].

"... SRC has led the way in research and development initiatives recognized across the province and around the world ... I am confident that the SRC will lead the way into that green and prosperous future we all dream of."

Premier Lorne Calvert, 2005 [170].

"... I am pleased to congratulate the Saskatchewan Research Council ... as one of Canada's Outstanding Employers ...I applaud your organization's long-standing commitment to our economic prosperity and community responsibility."

Premier Brad Wall, 2015 [171].

9 APPENDICES

LAURIER L. SCHRAMM

Appendix 9.1.   Canada's Provincial Research Councils.

References [15,22,172].

Province and Year(s) Established	Evolution
British Columbia Research Council, 1944	Privatized as BC Research Inc. in 1993.
Advisory Council of Scientific and Industrial Research of Alberta, 1921	Later renamed Science and Industrial Research Council of Alberta (SIRCA), then Research Council of Alberta (RCA), then Alberta Research Council (ARC), then became part of **Alberta Innovates – Technology Futures**.
Research Council of Saskatchewan (RCS), 1930; 1947	Re-established as the **Saskatchewan Research Council** (SRC) in 1947.
Manitoba Research Council, 1963	Evolved into the **Industrial Technology Centre** (ITC, 1979) and what is now the **Food Development Centre** (1978).
Ontario Research Foundation (ORF), 1928	A separate Ontario Research Council existed from the mid-1940s through 1955. ORF was renamed ORTECH Corp. in the 1990s, privatized around 1999; part was sold to Body Cote, part became Process Research ORTECH Inc.
Centre de Recherches Industrielle du Québec (CRIQ), 1969	
New Brunswick Research and Productivity Council, 1962	Later renamed the **Research and Productivity Council** (RPC).
PEI Food Technology Centre, 1987	Evolved into **Bio | Food | Tech** in 2011.
Nova Scotia Research Foundation Corp. (NSRFC)	Amalgamated into INNOVAcorp., a venture capital firm, in 1994.
Newfoundland Research Council	1961 NF Research Council Act was passed but not implemented. Research and Development Council formed under new Act in 2009, later renamed **Research and Development Corp.** (RDC).

LAURIER L. SCHRAMM

Appendix 9.2. An SRC Chronology.

1947 – 1955, The Early Years at SRC
1956 – 1972, The Building Years at SRC
1972 – 1983, The Maturing Years at SRC
1983 – 2000, The Commercial Years at SRC
2001 – 2017, The New Millennium Years at SRC

Year(s)	
1916	First research and technology organizations in North America, the U.S. NRC, and the Canadian NRC were launched
1930	First attempt to start SRC
1947	Successful re-launch of SRC
1947 - 1955	The Early Years at SRC
1948	NRC opens Prairie Regional Laboratory in Saskatchewan
1953	First full-time employee
1954	Technical Information Service (TIS) program in Saskatchewan
1956 - 1972	The Building Years at SRC
1956	First full-time CEO
1956	Industrial Minerals Research Branch acquired
1958	First major laboratory facilities opened
1963	U of S's historical climate records acquired
1963	First Climatological Reference Station launched
1969	Slurry pipeline facilities become world-class
1972 - 1983	The Maturing Years at SRC
1977	Saskatchewan Conservation House
1980/81	New major laboratory facilities opened in Resources Research Facility and SEDCO Centre
1981	SLOWPOKE-2 nuclear reactor commissioned and goes critical
1982	Canadian Centre for Advanced Instrumentation launched

Year(s)	
1983 - 2000	The Commercial Years at SRC
1983/84	Saskatchewan CAD/CAM Robotics Centre launched
1985	Petroleum R&D Centre acquired
Circa 1986	First natural gas and Dual-Fuel™ natural gas vehicles
1987	Bovine Blood Typing Laboratory acquired
1987 - 1995	Temporarily managed the Prairie Agricultural Machinery Institute (PAMI)
1990	Building Science Division acquired
Circa 1990	Environmental analytical laboratories become largest and best equipped in Canada
1993	Saskatchewan Advanced House
1995	Technology commercialization program created
1996	TecMark International Commercialization incorporated
1997/98	Fermentation pilot plant
1998	Strategic alliance in Petroleum Technology Research Centre
2001 - 2017	The New Millennium Years at SRC
2002	Office of Energy Conservation launched
2003	Collaboration with Mistawasis First Nation on groundwater resource evaluation and development launched
2004	Strategic drive to build a "world-class" safety culture launched
2004	World-first Dual-Fuel™ hydrogen vehicle demonstrated
Circa 2005	Uranium and diamond geoassay labs become world leading
2006	Collaboration with Federation of Saskatchewan Indian Nations on water quality in First Nations communities launched
2006	Contracted to manage remediation of abandoned cold war-era uranium mines in northern Saskatchewan (Project CLEANS)
2006	Biofuels Test Centre™ launched
Circa 2006	Potash geoassay laboratory becomes world-leading
2006/07	Agency Chiefs Centennial Home and Factor 9 Home launched

Year(s)	
2009	Saskatchewan Forest Centre programs acquired
2009	Advanced Microanalysis Centre™ launched
2009	World-class 3D high-pressure scaled physical model for enhanced oil recovery process development commissioned
2010	Collaboration with Conservation Learning Centre on boreal forest field laboratory launched
2010	Hydrogen loading/fuelling station launched with SaskEnergy
2010	Partnership with Cowessess First Nation in commercial demonstration high-level wind energy generation and storage
2011	2nd Climate Reference Station launched (near Prince Albert)
2012	Annual "*SRC Innovation Forum*" program launched
2015	Multi-purpose Shook-Gillies High Pressure, High Temperature Pipe Flow Facility for volatile/hazardous fluids launched
2016	Company-wide ISO 9001:2008 accreditation

LAURIER L. SCHRAMM

Appendix 9.3. Members of the SRC Board of Directors.

Member	Years Served
Anguish, Hon. D.	1992 - 1995
Archer, Dr. J.H.	1970 - 1975
Austin, Mr. E.	1961 - 1964
Begg, Dr.R.W.	1967 - 1970
Bennett, Mr. J.	2000 - 2008
Birney, Mr. G.	1977 - 1983; Chair 1983 - 1993
Black, Mr. D.H.F.	1953 - 1962
Blachford, Dr. C.	1981 - 1993
Bock, Mr. D.	Secretary 1974 - 1977
Bode, Ms. K.	2014 - Present
Boden, Mr. E.A.	1964 - 1967
Boileau, Mr. L.A.	1961 - 1964
Bonham-Smith, Dr. P.	2002-11; Vice-Chair 2011-14
Bracken, Dr. D.	1991 - 1993
Bratvold, Mr. D.	1990 - 1995
Brockelbank, Mr. E.E.	1947 - 1959
Bromberger, Mr. N.	1974 - 1977
Bryant, C.	1980
Burton, Mr. T.S.	1978 - 1981
Cameron, Mr. G.W.	1965 - 1970
Campbell, Mr. E.B.	1974 - 1983
Carter, Mr. N.	1980 - 1981
Cass-Beggs, Mr. D.	1958 - 1964
Cawley, Mr. J.T.	1958 - 1961; 1964 - 1970; Vice-Chair 1960
Cercone, Dr. N.	1993 - 1997
Chakma, Dr. A.	1997 - 2000
Colborne, Mr. G.L.	1967 - 1970
Cook, Ms. P.	2011 - Present
Cooper, Mr. L.	2000 - 2006
Corcoran, Dr. M.	1997 - 2000
Craig, B.M.	1974 - 1983
Cross, Mr. J.	1986-1989, 2009-Pres.; Vice-Chair 1989-1993, 1999-2000; Chair 1993-1999
Curran, Mr. A.	1983 - 1986
Currell, Mr. N.C.	1961 - 1968

Member (cont'd)	Years Served (cont'd)
Currie, Dr. B.W.	1973
Currie, Hon. G.J.	Chair 1983; Member 1983-1986
Dalgleish, Mr. R.G.	1975 - 1976
Degelman, Mr. W.	1980 - 1982
de la Gorgendiere, Mr. M.	1983 - 1990
Draude, Ms. J.	Vice-Chair 1993 - 1995
Eisenhauer, Mr. E.E.	Vice-Chair 1947 - 1953
Erb, Hon. J.W.	Chair 1961 - 1962
Fitzpatrick, Dr. D.	2011 - 2014; Chair 2014 - Present
Fletcher, Mr. J.E.	1974 - 1978
Foster, Mr. J.R.	1973 - 1974
Fraser, Dr. I.M.	1953 - 1958
Gartner, Dr. G.J.	1973; Vice-Chair 1974 - 1981
Gibbings, Mr. C.W.	1956 - 1961
Gilchrist, Ms. P.	2002 - 2010
Graham, Dr. V.E.	1947 - 1964
Grieve, Mr. T.	1999 - 2000
Guy, Mr. A.J.Y.	1982 - 1983; Vice-Chair 1983 - 1986
Hanson, Mr. K.	2000 - 2001; Chair 2001 - 2006
Hanson, Mr. O.	1983 - 1991
Haslam, Dr. R.N.H.	1971 - 1974
Hawkins, Mr. R.	1993 - 95; Vice-Chair 1995 - 99; Chair 1999 - 2000
Hesseln, Dr. H.	2011
Hogg, Mr. C.A.L.	1947 - 1974
Horner, Mr. W.H.	1960; Vice-Chair 1961 - 1974
Hutch, Mr. J.	1977 - 79; Vice-Chair 1983; Secretary 1983 - 95; Chair 2000 - 2001
Hutcheon, Dr. W.L.	1964 - 1965
Hutchinson, Mr. J.	Secretary 1998-2000
Jamieson, Dr. P.R.	2012 - 2014
Johnson, Dr. D.	1991 - 1997
Jonsson, Mr. J.O.	1980
Katz, Dr. L.	1975 - 1988
Kavadas, Mr. A.	1974 - 1978
Kelln, Mr. D.	Secretary 2000 - 2001; Member 2001 - 2007; Vice-Chair 2007 - 2010
Kirk, Dr. L.E.	1947 - 1948
Kobayashi-Keith, Ms. D.	1990 - 1993

Member (cont'd)	**Years Served** (cont'd)
Kononoff, Mr.V.	1990 - 2000
Kozinski, Dr. J.	2014 - Present
Kunkel, Mr. C.F.	1988 - 1991
Lautermilch, Hon. E.	1995 - 2000
Lepage, Mr. R.	2011 - 2014
Levin, Mr. D.	Secretary 1961 - 1963
Lipon, Ms. S.	2007 - 2014
Lloyd, Mr. R.E.	1975 - 1979
Lloyd, Hon. W.S.	Chair, 1947 - 1961
Lake, Mr. H.E.	1961 - 1970
Lokken, Mr. R.	1996 - 2000
Longworthy, Mr. W.O.	1956 - 1970
Lowe, Mr. A.	1993 - 1999
MacNicol, Mr. J.M.	1961 - 1967
MacMurchy, Hon. G.	Chair 1971 - 1975
Maliepaard, Mr. H.S.	1982 - 1988
Mann, Mr. T.	1989 - 1993
Manns, Dr. J.	1986 - 1991
Mantle, Mr. J.B.	1969 - 1974
Mawdsley, Dr. J.B.	1961 - 1964
McArdle, Mr. L.	1970 - 1971
McArthur, Mr. D.F.	1974 - 1975
McCallum, Dr. K.J.	1974 - 1984
McIntyre, Mr.. J.	2011 - 2014
McIsaac, Hon. J.C.	Chair 1967 - 1971
McKay, Mr. R.	1999 - 2000
McLeod, Dr. T.H.	1947 - 1953
McQuarrie, Mr. R.	1986 - 1989
Meiklejohn, Hon. R.	1986 - 1989
Melin, Ms. J.	1996 - 2000
Melin, Mr. P.	1978 - 1984
Messer, Hon. J.	Chair 1975 - 1976
Michel, Mr. B.	1978 - 1983
Missens, Ms. B.	1993 - 1996
Mitchell, Mr. G.C.	1967 - 1980
Mitura, Ms. V.	1993 - 2000
Monea, Mr. M.	Vice-Chair 2000 - 2006
Moore, Dr. J.J.	Secretary 1964 - 1969
Morris, Ms. A.	1983 - 1986

Member (cont'd)	Years Served (cont'd)
Murray, Mr. D.	1975 - 1976; Secretary 1977 - 1983
Nightingale, Mr. J.	1983 - 1986; 1988 - 89; Vice-Chair 1986 - 1988
Nikiforuk, Dr. P.N.	1978 - 2000
Olsen, Mr. H.D.	1988 - 1991
Olson, Mr. E.R.	1967 - 1974
Palidwar, Ms. K.	2007 - 2010
Patriquin, Mr. D.	1981; Vice-Chair 1982
Penner, Hon. J.	1991 - 1992
Perkins, Mr. J.	1980 - 1983
Phillips, Mr. R.H.D.	1967 - 1974
Pichler, Mr. D.	1974 - 1977
Porter, Dr. A.	1958 - 1961
Prefontaine, Mr. H.A.	1972 - 1973
Rak, Ms. D.	1988 - 1991
Randell, Dr. C.	2012 - 2014
Rhodes, Mr. N.	2012 - Present
Riddell, Dr. W.A.	1947 - 1969
Saddlemyer, Mr. K.O.	1973 - 1975
Sandberg, Hon. J.	Chair 1982 - 1983
Schramm, Dr. L.L.	Secretary 2001 - Present
Semple, Mr. G.	1986 - 1993
Setter, Mr. W.K.	Secretary, 1969 - 1974
Sharp, Mr. J.W.	1958 - 1960
Shoyama, K.T.	Secretary 1947 - 1961; Member 1962 - 1964
Sinclair, Mr. J.	1980 - 1981
Skelton, Ms. C.	1990 - 1993
Spencer, Mr. J.E.	1976 - 1979
Spencer, Dr. R.A.	1947 - 1955
Spinks, Dr. J.W.T.	1951 - 1974; 1983 - 1986
Steck, Dr. W.	1983 - 1986
Stevenson, Ms. S.	1986 - 1990
Stewart, Dr. C.D.	1965 - 1967
Sufrin, Mr. B.	1963 -1964
Swenson, Hon. R.	1989 - 1991
Teichrob, Ms. C.	1983 - 1990
Tennant, Dr. H.E.	1984 - 1986
Thompson, Mr. B.	1978 - 1983

Member (cont'd)	Years Served (cont'd)
Thorvaldson, Dr. T.	1947 - 1965
Tinker, Dr. E.B.	1976 - 1982
Tomkins, Mr. R.V.	1969 - 1974
Tosney, Mr. J.	1999 - 2000
Trapp, Hon. G.J.	Chair 1964 - 1966
Trask, Ms. P.	1993 - 2000
Turnbull, Hon. O.A.	Chair 1962 - 1964
Van Cleave, Dr. A.B.	1967 - 1974
Veawab, Dr. A.	2002 - 2005
Vickar, Hon. N.	Chair 1976 - 1982
Wallace, Mr. M.	1979 - 1980
Webster, Mr. C.R.	1965 - 1974
Weir, Ms. P.	1986 - 1988
Whelan, Mr. W.L.	1984- 1991
Williams, Mr. A.J.	1961 - 1964
Wilson, Dr. L.	2007 - 2010
Woodward, Mr. R.	Secretary 1995 - 1998
Wotherspoon, Mr. J.G.	1970 - 1974
Zawada, Mr. C.	2000 - 2007; Chair 2007 - 2013

Total number: 165

Mean years in position: 6.2

Longest serving: Mr. C.A.L. Hogg and Dr. John W.T. Spinks (28 years)

Dr. Dennis Fitzpatrick,
Board Chair

John Cross

Dr. Laurier Schramm,
President and CEO / Board Secretary

Dr. Janusz Kozinski,
Board Vice-Chair

Kelly Bode

Nathan Rhodes

Patricia Cook

Figure 9.1. SRC's Board of Directors in 2016 [141].

Appendix 9.4. Permanent Heads and Executive Team

Permanent Head	Years Served
Dr. Tom E. Warren	Director 1956 - 1972
Dr. Tom P. Pepper	President 1972 – 1983
Mr. Jim Hutch	President 1983 – 1995
Mr. Ron Woodward	President & CEO 1995 – 1998
Mr. Jim Hutchinson	President & CEO 1998 – 2000
Dr. Laurier L. Schramm	President & CEO 2001 – Present

Total number: 6

Mean years in position: 10.2

Longest serving: Dr. Warren (16.4 years)

Dr. Laurier Schramm,
President and CEO

Dr. Joe Muldoon,
VP, Environment

Toby Arnold,
VP, Organizational Effectiveness

Craig Murray,
VP, Mining and Minerals

Michael Crabtree,
VP, Energy

Wanda Nyirfa,
VP, Business Ventures

Ryan Hill,
VP, Finance

Phillip Stephan,
VP, Strategic Initiatives

Figure 9.2. SRC's Executive Team in 2016 [141].

Appendix 9.5. Distinguished Scientists and Researchers Emeriti.

Distinguished Scientist or Engineer. The designation of Distinguished Scientist or Engineer represents the highest position on SRC's Technical (non-management) "Ladder," and as such recognizes and emphasizes the technical accomplishments of the employee. A Distinguished Scientist or Engineer has a solid record of applying engineering or scientific solutions resulting in a direct effect on SRC's revenues and financial performance, reputation, and impacts.

Once appointed, the Distinguished Engineer/Scientist spends the majority of his/her time conducting research, development demonstration and/or deployment, developing business and coaching and leading other scientific, engineering or technical employees. The Distinguished Scientist or Engineer continues to contribute to strategic direction and realization of SRC's mission through the application of personal credentials, reputation and leadership capabilities.

Researcher Emeriti. The designation of Researcher Emeritus allows renowned former employees of SRC to continue, after retirement, working in areas of research where both the individual and SRC believe there are mutual benefits to the individual, SRC, and the province. A Division may sponsor a researcher (scientist or engineer), upon retirement, who has been employed by SRC for a significant period of time, as a Researcher/Scientist/Engineer Emeritus. Appointments are made by SRC's Executive Team.

Distinguished Scientists and Researchers Emeriti

Name	Field	
Dr. Randy Gillies, Distinguished Engineer (2005-2012), Researcher Emeritus (2012-Present)	Chemical engineering, pipeline hydrotransport, multi-phase flow, pipeline and centrifugal pump testing.	
Dr. Mark Johnston, Distinguished Scientist (2006-Present)	Forest resources management, forest ecology, climate change vulnerability, carbon accounting and GHG management, agroforestry.	
Dr. Elaine Wheaton, Distinguished Scientist (2008-2014), Researcher Emeritus (2011-Present)	Physical geography, climatology, global warming, climate change impacts and adaptation, drought assessments.	
Dr. Tony Kaminski, Researcher Emeritus (2009-Present)	Mechanical and agricultural engineering, industrial engineering, development engineering.	
Mike Sulatisky, Researcher Emeritus (2015-Present)	Mechanical engineering, development engineering, alternative energy, and fuel systems.	

Distinguished Scientists and Researchers Emeriti (cont'd)

Name	Field	
Dr. Jeff Thorpe, Researcher Emeritus (2014-Present)	Plant and forest ecology, grassland ecology, biodiversity, climate change impacts and adaptation.	
Keith Hutchence, Researcher Emeritus (2010–2015)	Chemistry, computer simulation of enhanced oil recovery, ethanol production and use, geothermal well systems.	
Dr. Ewen Coxworth, Researcher Emeritus (2003–2012)	Chemistry, production and processing of agricultural materials, agroforestry, and biofuels.	
Dr. Stan Shewchuk, Researcher Emeritus (2007–2013)	Physics, atmospheric sciences, climatology, meteorology, and air pollution.	

LAURIER L. SCHRAMM

Appendix 9.6. Acquisitions.

Year	Program or Organization	From	References
1949	Sask. contributions to the Prairie Rural Housing Committee	Sask. Department of Reconstruction and Rehabilitation	[27]
1954	Technical Information Service (TIS) program in Saskatchewan	National Research Council	[31]
1956	Industrial Minerals Research Branch	Sask. Department of Mineral Resources	[33]
1963	Climatological Reference Station	University of Saskatchewan	[40,41]
1985	(Petroleum) R&D Centre	SaskOil Corp.	[68]
1987	Bovine Blood Typing Laboratory	Ottawa	[75]
1990	Building Science Division	National Research Council, Ottawa	[113]
2002	Office of Energy Conservation (Start-Up)	Province of Saskatchewan, Regina	[116]
2009	Forest First Programs	Forest First, Prince Albert	
2012	Crop Evaluation (Grain-Testing) Laboratory	Canadian Wheat Board	

LAURIER L. SCHRAMM

Appendix 9.7. Spin-Offs and Spin-Outs from SRC.

Year	Program	To	References
1971	SRC Graduate Research Scholarships	Transferred to Provincial Government Scholarship Program[62]	[48,50]
1990	Archaeology Program	Privatized as Western Heritage Services Inc.	[173]
1995	Highway Test Track	Transferred to Canadian Transportation Innovation Centre	[64]
1995	Prairie Agricultural Machinery Institute[63] (PAMI)	Restored management autonomy to PAMI.	[67,73,78]
2005	Provincial Groundwater Observation Well Network and related programs	Saskatchewan Watershed Authority (SWA)	n/a
2014	Animal Genetics (formerly part of the Bova-Can Joint Venture)	Quantum Genetix Canada Inc.	n/a
2014	Crop Evaluation (Grain-Testing) Laboratory	Inspectorate Exploration & Mining Services	n/a

---

[62] SRC continued to conduct the selections of the scholarships until about 1988.

[63] PAMI was on the verge of discontinuance in 1987. SRC developed and implemented a five-year strategic plan for PAMI in 1988, and managed PAMI under contract from 1988 through 1995.

LAURIER L. SCHRAMM

Appendix 9.8. Books Published by SRC Staff.

With a strong focus on helping business, industry, and communities grow and thrive, and because so much of SRC's work is done under proprietary contracts, opportunities for SRC employees to publish their work are much reduced compared with the academic, government, or even the industrial sectors. However, quite a large number of technical presentations and publications have been given by SRC employees over the past 70 years, too many in fact to list here. The following list of books published by SRC employees provides some examples of areas in which SRC has demonstrated an unusually high degree of technical depth.

o Christiansen, Earl A. (Ed.), *Physical Environment of Saskatoon, Canada*, National Research Council of Canada, Ottawa, ON, 1970.
o Wheaton, Elaine, *But It's A Dry Cold: Weathering the Canadian Prairies*, Fifth House Publishers, Markham, ON, 1998.
o Schramm, Laurier L., *Emulsions, Foams, and Suspensions: Fundamentals and Applications*, Wiley - VCH, Weinheim, Germany, 2005.
o Schramm, Laurier L., *Dictionary of Nanotechnology, Colloid and Interface Science*, Wiley - VCH, Weinheim, Germany, 2008.
o Schramm, Laurier L. (Ed.), *Surfactants: Fundamentals and Applications in the Petroleum Industry*, Paperback Ed., Cambridge University Press, Cambridge, UK, 2010.
o Schramm, Laurier L., *Emulsions, Foams, Suspensions, and Aerosols: Microscience and Applications*, Wiley - VCH, Weinheim, Germany, 2014.
o Schramm, Laurier L., *Nano- and Microtechnology from A - Z: From Nanosystems to Colloids and Interfaces*, Wiley - VCH, Weinheim, Germany, 2014.
o Zhang, Jack; Zhao, Baodong; Schreiner, Bryan, *Separation Hydrometallurgy of Rare Earth Elements*, Springer, Heidelberg, 2016.
o Schramm, Laurier L., *Canada's Cold War Ghost Town, The Gunnar Uranium Mine*, Saskatchewan Research Council, Saskatoon, and Amazon.com Inc., 2016.
o Schramm, Laurier L., *Research and Development on the Prairies: A History of the Saskatchewan Research Council*, Saskatchewan Research Council, Saskatoon, and Amazon.com Inc., 2016.
o Schramm, Laurier L., *Technological Innovation: A Dictionary*, de Gruyter, Berlin, Germany, in press for 2017.
o Schramm, Laurier L., *Technological Innovation: An Introduction*, de Gruyter, Berlin, Germany, in preparation for 2017.

LAURIER L. SCHRAMM

Appendix 9.9. Patents Issued on Staff Inventions.

As is the case with technical presentations, publications, and books (Appendix 9.7), with such a strong focus on helping business, industry, and communities grow and thrive, and because so much of SRC's work is done under proprietary contracts, opportunities for SRC employees to get their inventions patented are much reduced compared with the academic, government, or even the industrial sectors. In many cases, intellectual property (IP) rights from SRC's work vest with the clients, and in quite a few cases any new IP is protected as trade secrets. The following list of issued patents with SRC inventors provides some examples of areas in which SRC has demonstrated an unusualy high degree of inventive skill.

The complete patents are normally available for free from the issuing countries. For Canadian and U.S. patents see:

- o http://www.ic.gc.ca/opic-cipo/cpd/eng/
- o http://patft.uspto.gov/

1980s:
- o Ranganathan, R., Pruden, B., "Simultaneous Hydrocracking of Bitumen/Coal Slurries," Canadian Patent 1,117,886, February 9, 1982.
- o Parker, R., Ranganathan, R., Lakshmanan, V., Last, A., "Method of Breaking a Water-In-Oil Emulsion," Canadian Patent 1,223,791, July 7, 1987.
- o McAndless, J.M., Hancock, J.R., Barnett, D., Olm, O., Locke, T., Maybank, J., "Polymer Packed Mini-Tube Vapour Sampling System," United States Patent 4,869,117, September 29, 1989.

1990s:
- o Englot, B., Olm, O., "Rechargeable Flashlight," United States Patent 4,903,178, February 20, 1990.
- o Phillips, W.C., Germann, A., Olm, O., "Compaction of Aluminum Beverage Cans," United States Patent 4,976,196, December 11, 1990.
- o McAndless, J.M., Hancock, J.R., Barnett, D., Olm, O., Locke, T., Maybank, J., "Automated Thermal Desorption Unit, United States Patent 4,976,924, December 11, 1990.
- o Haffey, D.J., Haffey, S.C., Olm, O., "Seat Belt Release Guard," United States Patent 4,987,662, January 29, 1991.

o Craig, W., Soveran, D., Production of Hydrocarbons with a Relatively High Cetane Rating," United States Patent 4,992,605, February 12, 1991.

o Craig, W., Soveran, D., "Production of Hydrocarbons with A Relatively High Cetane Rating," Canadian Patent 1,313,200, January 26, 1993.

o Scharf, J.E., Olm, O., "Plastic Wrap Dispenser," United States Patent 5,186,376, February 16, 1993.

o Englot, B., Olm, O., "Rechargeable Flashlight," Canadian Patent 1,315,757, April 4, 1993.

o McAndless, J.M., Hancock, J.R., Barnett, D., Olm, O., Locke, T., Maybank, J., "Automated Thermal Desorbtion Unit," Canadian Patent 1,327,282, March 1, 1994.

o McAndless, J.M., Hancock, J.R., Barnett, D., Olm, O., Locke, T., Maybank, J., "Polymer Packed Mini-Tube Vapour Sampling System," Canadian Patent 1,334,054, January 24, 1995.

o Davis, W.J., Olm, O., Seymore, B., "Pontoons and Pontoon Vessel," United States Patent 5,540,169, July 30, 1996.

o Monnier, J., Tourigny, G., Soveran, D., Wong, A., Hogan, D., Stumborg, M., "Conversion of Biomass Feedstock to Diesel Fuel Additive," United States Patent 5,705,722, January 6, 1998.

o Shabbits, G., Olm, O., "Dispensing Machine," United States Patent 5,794,820, August 18, 1998.

o Monnier, J.; Tourigny, G.; Soveran, D.W., "Conversion of Depitched Tall Oil to Diesel Fuel Additive," Canadian Patent 2,149,685, September 14, 1999.

o Sulatisky, M., White, N., "Electronic Gas Regulator," United Kingdom Patent 2,316,773, September 29, 1999.

o Sulatisky, M., White, N., "Electronic Gas Regulator," United States Patent 6,003,543, December 21, 1999.

2000s:

o Jaworski, E., Zlipko, D., "Water Level Monitoring Float System," Canadian Patent 2,229,396, October 3, 2000.

o Jaworski, E., Zlipko, D., "Water Level Monitoring Float System," United States Patent 6,138,509, October 31, 2000.

o Schmidt, B.H., Jespersen, P.J., Kristoff, B., "Multiple Drain Method for Recovering Oil from Tar Sand," United States Patent 6,263,965, July 24, 2001.

o Hill, S., Sulatisky, M., "High Volume Electronic Gas Regulator," Patent Cooperation Treaty (PCT) Patent WO0159537A1, August 16, 2001.

o Schmidt, B.H., Jespersen, P.J., Kristoff, B., "Multiple Drain Method for Recovering Oil from Tar Sand," Trinidad and Tobago Patent TT/P/2003/ 00001, February 11, 2003.

o Schmidt, B.H., Jespersen, P.J., Kristoff, B., "Multiple Drain Method for Recovering Oil from Tar Sand," Venezuela Patent 59.497, November 4, 2003.

o Sulatisky, M., Hill, S., Song, Y., Young, K., Gnanam, G., "Neural Control System and Method for Alternatively Fueled Engines," United States Patent 6,687,597, February 3, 2004.

o Sulatisky, M., Hill, S., "High Volume Electronic Gas Regulator," United States Patent 6,758,233, July 6, 2004.

o Lung, B., Wong, J.Y., "Method for Determining if Deterioration in Structural Integrity of a Pressure Vessel, a Pressure Vessel, and a Structural Integrity Testing Apparatus Therefor," United States Patent 6,785,616, August 31, 2004.

o Lung, B., Wong, J.Y., "Method for Determining if Deterioration in Structural Integrity of a Pressure Vessel, a Pressure Vessel, and a Structural Integrity Testing Apparatus Therefor," Canadian Patent 2,393,522, March 17, 2005.

o Sulatisky, M., White, N., "Electronic Gas Regulator," Canadian Patent 2,207,497, April 5, 2005.

o Sulatisky, M., Hill, S., "High Volume Electronic Gas Regulator," Canadian Patent 2,368,957, October 4, 2005.

o White, N., Lung, B., "System and Method for Monitoring and Controlling Gaseous Fuel Storage Systems," United States Patent 6,957,171, October 18, 2005.

o Schmidt, B.H., Jespersen, P.J., Kristoff, B., "Multiple Drain Method for Recovering Oil from Tar Sand," China Patent ZL99106916.1, January 4, 2006.

o Davis, J., Olm, O., Seymore, B., "Pontoons and Pontoon Vessel," Canadian Patent 2,089,058, November 7, 2006.

o Schmidt, B.H., Jespersen, P.J., Kristoff, B., "A Thermal Method for Recovering Hydrocarbon Oil," India Patent 215,675, February 29, 2008.

o Schmidt, B.H., Jespersen, P.J., Kristoff, B., "Multiple Drain Method for Recovering Oil from Tar Sand," Canadian Patent 2,272,593, September 30, 2008.

o Sulatisky, M., Hill, S., Song, Y., Young, K., Gnanam, G., "Neural Control System and Method for Alternatively Fueled Engines," Canadian Patent 2,422,970, May 26, 2009.

o   Reaney, M.J.T., Shen, J., Soveran, D.W., "Process for the Production of Polyol Base Catalysts," Canadian Patent application 2,707,101, June 4, 2009.

o   White, N., Lung, B., "System and Method for Monitoring and Controlling Gaseous Fuel Storage Systems," Canadian Patent 2,425,851, October 13, 2009.

2010s:

o   Hill, S., Sulatisky, M., "High Volume Electronic Gas Regulator," European Patent (including France, Germany, Great Britain, Netherlands, Italy, Sweden) 1,264,228, October 13, 2010.

o   Young, K.., Grimes, J.F., Peter, N., "Liquid Level Measuring System and Method," Canadian Patent 2,720,325, May 4, 2012.

o   Young, K.., Wan, Q., Farber, A.R.D., Sulatisky, M., Peter, N., Hill, S., "Method and System for Powering an Otto Cycle Engine Using Gasoline and Compressed Natural Gas," Canadian Patent 2,742,011, July 17, 2012.

o   Reaney, M.J.T., Shen, J., Soveran, D.W., "Process for the Production of Polyol Base Catalysts," United States Patent 8,410,010, April 2, 2013.

o   Young, K.., Wan, Q., Farber, A.R.D., Sulatisky, M., Peter, N., Hill, S., "Method and System for Powering an Otto Cycle Engine Using Gasoline and Compressed Natural Gas," United States Patent 8,590,515, November 26, 2013.

o   Young, K.., Grimes, J.F., Peter, N., "Liquid Level Measuring System and Method," United States Patent 9,046,407, July 2, 2015.

o   Jonasson, R., "Hydrocarbon Recovery Composition, Method, and System," United States Provisional Patent application filed July 11, 2016, patent pending.

Appendix 9.10. Associations of Provincial Research Councils.

NIRAC. The Non-Profit Industrial Research Association of Canada (NIRAC) was formed in 1969 to promote recognition of the existence, the value and the potential of non-profit industrial research organizations in Canada and to ensure for them a place in any national science policy. Members were the Nova Scotia and Ontario Research Foundations, the Research Councils of Alberta and Saskatchewan, New Brunswick Research and Productivity Council, and B.C. Research. In 1969 the provincial research councils had available to them about 670 000 square feet of laboratory, office, pilot plant and other space, and about $8 million worth of major equipment items [22]. Collaborative activities among the industrial research organizations under NIRAC seem to have rapidly diminished after 1972.

APRO. The provincial research organizations (PROs) seem to have started actively meeting together again in about 1983, and later that year they started using the name Association of Provincial Research Organizations (APRO). This relationship was formalized by the incorporation of The Association of Provincial Research Organizations Inc. in November 1984, although the short name APRO continued to be used for most purposes. The original APRO membership comprised the research councils of BC, Alberta, Saskatchewan, Manitoba, Ontario, Québec, New Brunswick, and Nova Scotia.

APRO meetings were commonly held in Ottawa to facilitate discussions with federal representatives. The goal of APRO was to *"[use] science and technology for the economic and social development of Canada and the Provinces"* [174]. The APRO meetings gave the PROs *"an opportunity to exchange ideas among themselves and develop strategies, … raise the level of awareness of the PROs and their capabilities to outsiders, … [and] enable them to press their case jointly in connection with problems that may exist with policies or programs of federal departments …"* [15].

In 1991 APRO became "APRO – The Canadian Technology Network" [80], and in about 1996 APRO was amalgamated with the Canadian Manufacturers' Association (CMA) and the Canadian Exporters' Association (CEA) to form the Alliance of Manufacturers & Exporters Canada, which in 2000 became Canadian Manufacturers & Exporters (CME).

I-CAN. After another lapse of nearly a decade, four[64] of the provincial research organizations (PROs) started meeting together again in 2003. These four decided to re-create some kind of association, which for a short time was referred to as the Alliance of Provincial Research Organizations, and resurrecting the older acronym APRO. The aims of the new alliance would be to strengthen Canada's innovation performance as a means of maintaining and improving Canada's global competitiveness, and to help Canada's research and technology organizations (RTOs) come together to work on projects of regional and/or national interest that would enable industry to solve large technology problems and challenges that would be too broad, expensive or risky for any one individual RTO to address alone.

The Founding Directors were: John McDougall, Dr. Laurier Schramm, Dr. Denis Beaulieu, and Trevor Cornell, the first Members were ARC, SRC, ITC, and CRIQ, and the new alliance was named Innoventures Canada Inc. (I-CAN™). I-CAN was incorporated in 2006. By 2015 more PROs had joined I-CAN[65]: FP Innovations (BC and Quebec), the National Optics Institute (INO, Quebec), Bio|Food|Tech (PEI), New Brunswick Research and Productivity Centre (RPC), National Research Council (NRC), and Vineland Research and Innovation Centre (Ontario). This gave I-CAN coast-to-coast "reach" across Canada, with (collectively) over 6,000 employees, over $1 billion/yr in market-oriented R&D services, and over 150 facilities across all ten provinces.

One of I-CAN's flagship programs has been the *I-CAN™ Innovation School,* which was developed to address the needs for education, training, and continuous professional development in technological innovation, research and technology organizations (RTOs), and in the components and the workings of effective innovation systems [175]. The first annual *I-CAN™ Innovation School* was launched in 2010, and it has continued to the present day.

By 2011 I-CAN members had collectively stimulated over $1 billion in direct business investment in science and technology, created several billion dollars in economic benefits through technology commercialization, attracted significant foreign benefits, improved Canadian quality of life by mitigating or reducing environmental impacts, and provided scientific understanding for more effective government regulations [175].

---

[64] Alberta Research Council, Saskatchewan Research Council, Industrial Technology Centre, and Centre de Recherche Industrielle du Québec.

[65] The Northern Centre for Advanced Technology (NORCAT) and Powertech Labs also joined but later left the alliance.

Appendix 9.11. Glossary and Acronyms.

**Advisory Council
of Scientific and
Industrial Research
of Alberta** *See* Alberta Innovates – Technology Futures.

AI-TF *See* Alberta Innovates – Technology Futures.

Afforestation Growing trees on land that was not previously covered by trees. This is different from "reforestation," which is involves re-establishing trees on land that was previously covered by trees. *See also* Agroforestry.

Agroforestry The practice of growing trees adjacent to or within crops or pastures. In addition to helping to protect the crops and fields, agroforestry can provide an addition farm product – the trees themselves. *See also* Afforestation.

**Alberta Innovates –
Technology
Futures** (AI-TF) The first provincial research council to be established in Canada, it was created as the Advisory Council of Scientific and Industrial Research of Alberta in 1921, later renamed Science and Industrial Research Council of Alberta (SIRCA), then Research Council of Alberta (RCA), then Alberta Research Council (ARC), then Alberta Innovates – Technology Futures (AI-TF).

**Alberta Research
Council** (ARC) *See* Alberta Innovates – Technology Futures.

ARC Alberta Research Council. *See* Alberta Innovates – Technology Futures.

**BC Research
Inc**. *See* British Columbia Research Council.

Bio-Ethanol   Ethanol produced from biological materials. The principal commercial process for the production of ethanol involves microbial fermentation of the sugars in plants (starch and cellulose).

Bio | Food | Tech

The ninth "provincial research council" to be established in Canada. Prince Edward Island (PEI) did not create a research council *per se*, however, the PEI Food Technology Centre was incorporated in 1987 by the provincial government as a wholly-owned subsidiary of Innovation PEI, a Provincial Crown Corporation. Its name was changed to Bio | Food | Tech in 2011. In modern terminology, it is a research and technology organization (RTO) as are the provincial research councils.

British Columbia
Research
Council   The fourth provincial research council to be established in Canada, it was created in 1944. It was privatized as BC Research Inc. in 1993.

CAD/CAM
Robotics Centre (Computer-Aided Design, Computer-Aided Manufacturing Robotics Centre) An SRC "centre" launched in 1983/84 with the aim of assisting manufacturing clients with advanced product design and prototyping support. This program was operated for several years before being merged into the Product and Process Development Division in 1988.

Canadian Centre
for Advanced
Instrumentation (CCAI) An SRC "centre" launched in 1982 with the aim of providing advanced instrumentation design, development, and services to the natural resource and high-technology industries. In 1988 diminishing external funding led to CCAI being mostly wound down, with some of its activities being maintained, but re-integrated within other existing SRC Divisions.

CCAI   *See* Canadian Centre for Advanced Instrumentation.

**Centre de Recherches
Industrielle du
Québec**   (CRIQ) The eighth provincial research council to be established in Canada, it was created in 1969.

C-14   A specific isotope of carbon (C) that is radioactive (hence the term *"radiocarbon"*), whereas the most common isotopes of carbon (C-12 and C-13) are not radioactive. C-14 is created in earth's atmosphere and forms radioactive carbon dioxide which is ultimately taken-up by living plants and animals. Following the death of such plants or animals, the concentrations of C-14 in them decreases with time enabling the radiocarbon dating technique.

CNES   *Centre National d'Etudes Spatiales* (France).

CNG   Compressed natural gas. When used as an alternative fuel, compressed natural gas (CNG) is stored onboard in high-pressure tanks in which the gas has been compressed to about 200 bar (3,000 psi). CNG is different from LPG (liquefied natural gas) in that LNG is natural gas that has been supercooled to about -120 °C or less (-184 F or less) which converts it to liquid form.

CRIQ   *See* Centre de Recherches Industrielle du Québec.

CSR   Corporate Social Responsibility.

C-TIC   Canadian Transportation Innovation Centre.

DNC   Delayed neutron counting. A non-destructive SLOWPOKE reactor method for uranium assaying.

EOR   Enhanced oil recovery. This normally refers to a stage of oil recovery that follows primary drive mechanisms (like natural underground pressure) and subsequent waterflooding, and requires the use of additional drive mechanisms (like steam, solvents, or chemical solutions).

ERS-1　　　An ESA European remote sensing satellite, and the first Earth-observing satellite program to use a polar orbit.

ESA　　　European Space Agency.

GMOs　　　Genetically modified organisms.

Ground-Truthing

In remote sensing, "ground truthing" refers to collecting information "on location" as opposed to from a distance. The process of ground truthing is used to compare remotely acquired imagery with the actual features and objects on the ground in order to calibrate the imagery, to avoid the distortions inherent in long-range imaging, and/or as an aid to interpretation of the imagery.

**Honorary Advisory Council for
Scientific and Industrial
Research**　　　The original name for the National Research Council (NRC), which became better known under its newer name after 1925.

Horizontal Well　In the petroleum industry this refers to a well that has been drilled at any angle other than vertical, and it need not be horizontal. Drilling horizontal wells is also referred to as "directional drilling."

Ice Road　　　A roadway constructed on the naturally frozen winter surface of a lake, bay, or river in order to link land roads. Ice roads are necessarily temporary and are only safe for vehicular travel when the minimum ice thickness exceeds a critical value and given ice thicknesses can only support vehicles of certain mass and speed limits. Ice roads are often winter substitutes for summer ferry services. Also termed *"Ice Crossing"* or *"Ice Bridge."*

**Industrial Research
Assistance
Program**　　　(IRAP) *See* Technical Information Service.

**Industrial Technology
Centre**　　　*See* Manitoba Research Council.

INNOVAcorp   *See* Nova Scotia Research Foundation Corp.

IRAP   *See* Technical Information Service.

ITC   Industrial Technology Centre. *See* Manitoba Research Council.

LANDSAT   NASA's LANDSAT satellites have provided the longest-duration acquisition of satellite imagery of Earth in history.

LPG   Liquefied Natural Gas. *See* CNG.

MASS   *See* MINITUBE™ Air Sampling System.

**MINITUBE™
Air Sampling
System**   (MASS) An automated air sampling machine developed by SRC in 1983 and commercialized as a product in 1984.

**Manitoba Research
Council**   The seventh provincial research council to be established in Canada, it was created in 1963. It later evolved into the Industrial Technology Centre (ITC, 1979) and what is now the Food Development Centre (1978).

NAA   Neutron-activation analysis. A non-destructive SLOWPOKE reactor method for halogenated-organics analyses (among other things), for compounds like polychlorinated biphenyls PCBs.

NASA   National Aeronautics and Space Administration (U.S.).

**National Research Council
of Canada**   (NRC) Canada's national research and technology organization, formed in 1916.

**New Brunswick
Research and Productivity
Council**   *See* Research and Productivity Council.

NOAA   National Oceanic and Atmospheric Administration (U.S.).

NOAA-2 A NOAA satellite that was mainly used for cloud imagery and temperature profiling.

Nova Scotia Research Foundation
Corp. (NSRFC) The fifth provincial research council to be established in Canada, it was created in 1946. It was amalgamated into INNOVAcorp., a venture capital firm, in 1994.

NRC *See* National Research Council of Canada.

NRCan Natural Resources Canada.

NSRFC *See* Nova Scotia Research Foundation Corp.

OH&S Occupational health and safety.

Ontario Research
Foundation (ORF) The second provincial research council to be established in Canada, it was created as the Ontario Research Foundation in 1927. A separate Research Council of Ontario existed from the mid-1940s through 1955 after which its activities were transferred to ORF. ORF was renamed ORTECH Corp. in the 1990s and privatized around 1999.

ORF *See* Ontario Research Foundation.

ORTECH
Corp. *See* Ontario Research Foundation.

PAMI *See* Prairie Agricultural Machinery Institute.

PCBs Polychlorinated biphenyl compounds. *See* NAA.

PEI Food Technology
Centre *See* Bio | Food | Tech.

Prairie Agricultural Machinery

Institute   (PAMI) A Humboldt, Saskatchewan-based research and technology organization (RTO) focused on applied research, development, testing, and evaluation related to farm machinery. SRC managed PAMI under contract from 1988 through 1995, at which time it was spun (back) out as an autonomous organization.

PARC   Prairie Adaptation Research Cooperative.

PRO   Provincial Research Organization. Generally, either a provincial research council or a not-for-profit organization with a similar mandate, mission, and mode of operation to those of a provincial research council.

PTRC   Petroleum Technology Research Centre.

R&D   Research and development, this acronym is frequently meant to refer specifically to applied research and development.

Radiocarbon
Dating   *See* C-14.

RCA   Research Council of Alberta. *See* Alberta Innovates – Technology Futures.

RCS   Research Council of Saskatchewan. *See* Saskatchewan Research Council.

RDC   *See* Research and Development Corp.

Reforestation   *See* Afforestation.

Research and Development

Corp.   (RDC) The tenth provincial research council to be established in Canada, it was created in Newfoundland and Labrador as the Research and Development Council in 2009 and later renamed Research and Development Corp. (RDC).

Newfoundland Research

Council   There was never a "Newfoundland Research Council" *per se*, but see Research and Development Corp.

Research and Productivity

Council   (RPC) The sixth provincial research council to be established in Canada, it was created as the New Brunswick Research and Productivity Council in 1962 and later renamed Research and Productivity Council.

Research Council

of Alberta   (RCA) *See* Alberta Innovates – Technology Futures.

Research Council

of Ontario   *See* Ontario Research Foundation.

Research Council of

Saskatchewan   (RCS) *See* Saskatchewan Research Council.

RPC   See Research and Productivity Council.

Saskatchewan Research

Council   (SRC) The third provincial research council to be established in Canada, it was created as the Research Council of Saskatchewan (RCS) 1930, later collapsed, and later re-established as the Saskatchewan Research Council in 1947.

Science and Industrial Research Council

of Alberta   (SIRCA) *See* Alberta Innovates – Technology Futures.

SIR   Saskatchewan Industry and Resources (later to become the Saskatchewan Ministry of Economy).

SIRCA   Science and Industrial Research Council of Alberta. *See* Alberta Innovates – Technology Futures.

Small- and Medium-Sized

Enterprises (SMEs) Business enterprises that are smaller than a specified number of employees (e.g., 500), and/or have annual revenues of less than a specified value (e.g., $75 million), as distinguished from large-size enterprises.

SMEs *See* Small- and Medium-Sized Enterprises.

SPOT A CNES satellite program providing high-resolution optical-imaging Earth observation.

SRC *See* Saskatchewan Research Council.

Technical Information

Service (TIS) An NRC program aimed at assisting industry by making appropriate research and technical information available to it. This program was delivered (for NRC) in Saskatchewan by SRC beginning in 1954 and until it was superseded by NRC's broader Industrial Research Assistance Program (IRAP) Program in 1981. IRAP focuses on providing a range of technical services to small- and medium-sized enterprises (SMEs) in Canada.

TIS *See* Technical Information Service.

U of R University of Regina.

U of S University of Saskatchewan.

LAURIER L. SCHRAMM

10 SUMMARY

Research and Development on the Prairies.
A History of the Saskatchewan Research Council

Early in the 20th century, the advent of industrial research councils brought organized research, development, and technological innovation to North America. Such organizations, now usually referred to as research and technology organizations (RTOs), focused on research and development aimed at helping industry develop and advance, and they were critical to the evolution of the modern approach to technological innovation. One of Canada's first RTOs to be established was the Saskatchewan Research Council (SRC), and it has become one of the most enduring. This book traces the evolution of SRC from its first efforts in the 1930s, through several distinct eras comprising its re-start mode in the *Early Years (1947-1955)*, its *Building Years* (1956-1972), *Maturing Years* (1972-*1983*), *Commercial Years* (1983-2000), and *New Millennium Years* (2001-2017).

Print ISBN: 978-0-9958081-3-3
ePub ISBN: 978-0-9958081-1-9

LAURIER L. SCHRAMM

11 ABOUT THE AUTHOR

Dr. Laurier Schramm has over 35 years of R&D experience spanning each of the industry, not-for-profit, university, and government sectors. He is currently President and CEO of the Saskatchewan Research Council (SRC). His interests include technological innovation, management and leadership, colloid & interface science, and nanotechnology. He holds 17 patents and has published 13 books and over 400 other publications and proprietary reports. He has served on many expert advisory panels and Boards, is co-founder of Innoventures Canada Inc. (I-CAN), and co-founder of Canada's Innovation School™. He has received national scientific and engineering awards for his work and is a Fellow of the Chemical Institute of Canada and an honourary Member of the Engineering Institute of Canada.

LAURIER L. SCHRAMM

12 REFERENCES

1. Douglas, T.C., "Official Opening [of the SRC Building]," Brochure, Saskatchewan Research Council, 1 October 1958.
2. Spinks, J.W.T., "University Research and Research in Cooperation with the Saskatchewan Research Council to Date," Presented at *Symposium on Research* and Opening Ceremonies for SRC's new building, 1 October 1958.
3. Steacie, E.W.R., "The Government Laboratory and Canadian Research," Presented at *Symposium on Research* and Opening Ceremonies for SRC's new building, 1 October 1958, from Saskatchewan Research Council historical files.
4. Holland, M., *Industrial Explorers*, Harper & Bros.: New York, 1928.
5. Holland, M., "Research, Science and Invention," In *A Century of Industrial Progress*, Wile, F.W. (Ed.), Doubleday: N.Y., 1928, pp. 312-334.
6. Schramm, L.L., *Technological Innovation – An Introduction*, de Gruyter, Berlin, *in preparation for 2017*.
7. Thistle, M., *"The Inner Ring. The Early History of the National Research Council of Canada*, University of Toronto Press, Toronto, 1966.
8. Eggleston, W., *"National Research in Canada, The NRC 1916 – 1966,"*, Clarke, Irwin & Co., Toronto, 1978.
9. Phillipson, D.J.C., "The National Research Council of Canada: Its Historiography, its Chronology, its Bibliography," *Scientia Canadensis*, **1991**, *15(2)*, 177-193.
10. The Advisory Council of Scientific and Industrial Research of Alberta, "First Annual Report," King's Printer, Edmonton, 1921.
11. The Research Council of Alberta, "Thirty-First Annual Report," King's Printer, Edmonton, 1951.
12. The Research Council of Alberta, "Annual Report, 1943," King's Printer, Edmonton, 1945.

13. Research Council of Alberta, "Thirty-Sixth Annual Report," Queen's Printer, Edmonton, 1956.

14. Aronovitch, D. (Ed.), "Ontario Research Foundation," In *The Canadian Encyclopedia*, 2013, http://www.thecanadianencyclopedia.ca/en/article/ontario-research-foundation/.

15. Le Roy, D.J.; Dufour, P., "Partners in Industrial Strategy. The Special Role of the Provincial Research Organizations," Background Study for the Science Council of Canada, Special Study 51, Supply and Services Canada, Ottawa, Nov., 1983.

16. Research Council of Saskatchewan, "Annual Report, 1930," Research Council of Saskatchewan, Regina, 1932.

17. Research Council of Saskatchewan, "Annual Report, 1932," Research Council of Saskatchewan, Regina, 1933.

18. Warren, T.E., "A Brief History of the Saskatchewan Research Council. 1914 – 1972," Report, Saskatchewan Research Council, Saskatoon, 1972.

19. Saskatchewan Archives Board, *Personal Communication*, 24 November, 2011.

20. Aronovitch, D. (Ed.), "British Columbia Research Council," In *The Canadian Encyclopedia*, 2013, http://www.thecanadianencyclopedia.ca/en/article/british-columbia-research-council/.

21. Aronovitch, D. (Ed.), "Nova Scotia Research Foundation Corporation," In *The Canadian Encyclopedia*, 2013, http://www.thecanadianencyclopedia.ca/en/article/nova-scotia-research-foundation-corporation/.

22. Wilson, A.H., "Research Councils in the Provinces: A Canadian Resource," Background Study for the Science Council of Canada, Special Study 19, Information Canada, Ottawa, June, 1971.

23. Government of Saskatchewan, *"The Research Council Act,"* 1947, c.114; as amended by The Revised Statutes of Saskatchewan, 1978 (Supplement), c.60; and the Statutes of Saskatchewan, 1983-84, c.34; 1988-89, c.22; 1991, c.T-1.1; 1994, c.45; 2000, c.23; and 2014, c.E-13.1; Queen's Printer, Regina, 2014.

24. SRC, *Annual Report of the Saskatchewan Research Council 1947*, Saskatchewan Research Council, Regina, 1948.

25. SRC, *Fourth Annual Report of the Saskatchewan Research Council 1950*, Saskatchewan Research Council, Regina, 1951.

26. SRC, *Second Annual Report of the Saskatchewan Research Council 1948*, Saskatchewan Research Council, Regina, 1949.

27. SRC, *Third Annual Report of the Saskatchewan Research Council 1949*, Saskatchewan Research Council, Regina, 1950.

28. SRC, *Fifth Annual Report of the Saskatchewan Research Council 1951*, Saskatchewan Research Council, Regina, 1952.

29. SRC, *Sixth Annual Report of the Saskatchewan Research Council 1952*, Saskatchewan Research Council, Regina, 1953.
30. SRC, *Seventh Annual Report of the Saskatchewan Research Council 1953*, Saskatchewan Research Council, Regina, 1954.
31. SRC, *Eighth Annual Report of the Saskatchewan Research Council 1954*, Saskatchewan Research Council, Regina, 1955.
32. SRC, *Ninth Annual Report of the Saskatchewan Research Council 1955*, Saskatchewan Research Council, Regina, 1956.
33. SRC, *Tenth Annual Report of the Saskatchewan Research Council 1956*, Saskatchewan Research Council, Regina, 1957.
34. SRC, *1983 Annual Report, Saskatchewan Research Council*, Saskatchewan Research Council, Saskatoon, 1984.
35. SRC, *19th Annual Report of the Saskatchewan Research Council 1965*, Saskatchewan Research Council, Regina, 1966.
36. SRC, *26th Annual Report of the Saskatchewan Research Council 1972*, Saskatchewan Research Council, Saskatoon, 1973.
37. SRC, *11th Annual Report of the Saskatchewan Research Council 1957*, Saskatchewan Research Council, Regina, 1958.
38. SRC, *12th Annual Report of the Saskatchewan Research Council 1958*, Saskatchewan Research Council, Regina, 1959.
39. Industry and Commerce, "The Saskatchewan Research Council," *The Growth Province*, Special Supplement, Saskatchewan Department of Industry and Commerce, Regina, September-October, 1968, 6 pp.
40. SRC, *16th Annual Report of the Saskatchewan Research Council 1962*, Saskatchewan Research Council, Regina, 1963.
41. SRC, *17th Annual Report of the Saskatchewan Research Council 1963*, Saskatchewan Research Council, Regina, 1964.
42. SRC, *18th Annual Report of the Saskatchewan Research Council 1964*, Saskatchewan Research Council, Saskatoon, 1965.
43. SRC, *14th Annual Report of the Saskatchewan Research Council 1960*, Saskatchewan Research Council, Regina, 1961.
44. SRC, *13th Annual Report of the Saskatchewan Research Council 1959*, Saskatchewan Research Council, Regina, 1960.
45. SRC, *20th Annual Report of the Saskatchewan Research Council 1966*, Saskatchewan Research Council, Regina, 1967.
46. Bergsteinsson, J.L.; Taylor, W.E., "First 25 Years," Report, Saskatchewan Research Council, Saskatoon, 1972.
47. SRC, *23rd Annual Report of the Saskatchewan Research Council 1969*, Saskatchewan Research Council, Regina, 1970.
48. SRC, *25th Annual Report of the Saskatchewan Research Council 1971*, Saskatchewan Research Council, Regina, 1972.
49. SRC, *27th Annual Report of the Saskatchewan Research Council 1973*, Saskatchewan Research Council, Regina, 1974.
50. SRC, *24th Annual Report of the Saskatchewan Research Council 1970*, Saskatchewan Research Council, Regina, 1971.

51. SRC, *Forward from 50. Celebrating 50 Years of Applied Research & Development Services to Saskatchewan*, SRC Brochure, Saskatchewan Research Council, Saskatoon, 1997.

52. SRC, *28th Annual Report of the Saskatchewan Research Council 1974*, Saskatchewan Research Council, Saskatoon, 1975.

53. SRC, *33rd Annual Report of the Saskatchewan Research Council 1979*, Saskatchewan Research Council, Saskatoon, 1980.

54. SRC, 1980 *Annual Report*, Saskatchewan Research Council, Saskatoon, 1981.

55. SRC, *1983 Annual Report*, Saskatchewan Research Council, Saskatoon, 1984.

56. SRC, *32nd Annual Report of the Saskatchewan Research Council 1978*, Saskatchewan Research Council, Saskatoon, 1979.

57. SRC, *31st Annual Report, 1977*, Saskatchewan Research Council, Saskatoon, 1978.

58. SRC, *1981 Annual Report*, Saskatchewan Research Council, Saskatoon, 1982.

59. Hilborn, J.W.; Burbridge, G.A., "SLOWPOKE: The First Decade and Beyond," Report AECL-8252, Atomic Energy of Canada Ltd., Chalk River, ON, October, 1983.

60. SRC, *29th Annual Report, 1975*, Saskatchewan Research Council, Saskatoon, 1976.

61. Meyer, D., "Archaeology at the Saskatchewan Research Council: History, Role, and Contribution," *J. Sask. Archaeological Soc.*, **1987**, *8(1)*, *ca.* 24pp.

62. SRC, *30th Annual Report, 1976*, Saskatchewan Research Council, Saskatoon, 1977.

63. SRC, *1982 Annual Report*, Saskatchewan Research Council, Saskatoon, 1983.

64. SRC, *Annual Report 1995-1996*, Saskatchewan Research Council, Saskatoon, 1996.

65. SRC, *Annual Report 1993-1994*, Saskatchewan Research Council, Saskatoon, 1994.

66. SRC, 5 Year Strategic Plan," Confidential internal report, Saskatchewan Research Council, Saskatoon, 1992.

67. SRC, *Annual Report 1987-1988*, Saskatchewan Research Council, Saskatoon, 1989.

68. SRC, *Annual Report 1988-1989*, Saskatchewan Research Council, Saskatoon, 1989.

69. SRC, *Annual Report 1984-1985*, Saskatchewan Research Council, Saskatoon, 1985.

70. SRC, *Annual Report 1985-1986*, Saskatchewan Research Council, Saskatoon, 1986.

71. SRC, *Annual Report 1983-84*, Saskatchewan Research Council, Saskatoon, 1984.

72. Barnett, D., Olm, O., Locke, T., Maybank, J., "Polymer Packed Mini-Tube Vapour Sampling System," *United States Patent* 4,869,117, September 29, 1989.

73. SRC, *Annual Report 1994-1995*, Saskatchewan Research Council, Saskatoon, 1995.

74. SRC, *Annual Report 1996-1997*, Saskatchewan Research Council, Saskatoon, 1997.

75. SRC, *Annual Report 1986-1987*, Saskatchewan Research Council, Saskatoon, 1987.

76. SRC, *Annual Report 1989-1990*, Saskatchewan Research Council, Saskatoon, 1990.

77. SRC, *Annual Report 1992-1993*, Saskatchewan Research Council, Saskatoon, 1993.

78. SRC, *Annual Report 1998-1999*, Saskatchewan Research Council, Saskatoon, 1999.

79. SRC, *Annual Report 1990-1991*, Saskatchewan Research Council, Saskatoon, 1991.

80. SRC, *Annual Report 1991-1992*, Saskatchewan Research Council, Saskatoon, 1992.

81. Sun Ridge, "Saskatchewan Advanced House," *Saskatchewan Advanced House Magazine*, Sun Ridge Group, Saskatoon, January, 1993, pp. 3-22.

82. SRC, *Annual Report 1999-2000*, Saskatchewan Research Council, Saskatoon, 2000.

83. Soveran, D., Craig, W., Production of Hydrocarbons with a Relatively High Cetane Rating," *United States Patent* 4,992,605, February 12, 1991.

84. Soveran, D., "Production of Hydrocarbons with A Relatively High Cetane Rating," *Canadian Patent* 1,313,200, January 26, 1993.

85. Soveran, D., Monnier, J., Tourigny, G., Wong, A., Hogan, D., Stumborg, M., "Conversion of Biomass Feedstock to Diesel Fuel Additive," United States Patent 5,705,722, January 6, 1998.

86. Monnier, J.; Tourigny, G.; Soveran, D.W., "Conversion of Depitched Tall Oil to Diesel Fuel Additive," Canadian Patent 2,149,685, September 14, 1999.

87. SRC, *Annual Report 1997-1998*, Saskatchewan Research Council, Saskatoon, 1998.

88. SRC, *50th Anniversary Brochure*, SRC Pub. No. G751-2-G-96, Saskatchewan Research Council, Saskatoon, 1996.

89. SRC, *Annual Report 2000-2001*, Saskatchewan Research Council, Saskatoon, 2001.

90. Warick, J., "Gov't Fires SRC Board, President," *Regina LeaderPost*, October 6, 2000, p. B1.

91. Warick, J., "Gov't Tosses Out Entire SRC Board," *Saskatoon Star Phoenix*, October 6, 2000, pp. A1-A2.

92. Saskatchewan Hansard, *Verbatim Minutes*, Committee of Finance, June 18, 2001, Legislative Assembly of Saskatchewan, pp. 1894-1899.

93. Grier, D.; Mengu, M., "Best Practices for Management of Research & Technology Organizations," Proc. IAMOT Conference, Orlando, FL, International Association for Management of Technology, 1998.

94. SRC, *2001-2002 Annual Report*, Saskatchewan Research Council, Saskatoon, 2002.

95. SRC, *Smart Science Solutions. 2002-2003 Annual Report*, Saskatchewan Research Council, Saskatoon, 2003.

96. SRC, *Impacting Your World. 2003-2004 Annual Report*, Saskatchewan Research Council, Saskatoon, 2004.

97. SRC, *Looking to the Future. 2006/2007 Annual Report*, Saskatchewan Research Council, Saskatoon, 2007.

98. SRC, *Ready, Set, Grow. 2009-2010 Annual Report*, Saskatchewan Research Council, Saskatoon, 2010.

99. SRC, *Responsible Science Solutions. 2007-2008 Annual Report*, Saskatchewan Research Council, Saskatoon, 2008.

100. SRC, *Science Serving Saskatchewan. 2004/2005 Annual Report*, Saskatchewan Research Council, Saskatoon, 2005.

101. Jansen, R.; Rohraff, D., "Case Study: Powering a Remote Remediation Camp with Diesel, Renewables and Energy Storage," *Energy and Mines* (online), October 23, 2015 (http://energyandmines.com/2015/10/case-study-powering-a-remote-remediation-camp-with-diesel-renewables-and-energy-storage/). Also available at http://blog.src.sk.ca/energy/case-study-powering-a-remote-remediation-camp-with-hybrid-energy/.

102. Saskatchewan, "North America's Newest Biofuel Test Centre Opens," News Release, Government of Saskatchewan, Regina, 21 September, 2006, http://www.saskatchewan.ca/government/news-and-media/2006/september/21/north--americas--newest--biofuel-test-centre-opens.

103. Western Diversification, "Governments of Canada and Saskatchewan support Mining Industry Research Lab," News Release, Western Economic Diversification Canada, 2 October, 2009, http://www.wd.gc.ca/eng/77_11619.asp.

104. SRC, *Sustaining Growth. 2010/2011 Annual Report*, Saskatchewan Research Council, Saskatoon, 2011.

105. SRC, *Sustainable Science Solutions. 2005/2006 Annual Report*, Saskatchewan Research Council, Saskatoon, 2006.

106. SRC, *Exploring the Next Frontier. 2008-2009 Annual Report*, Saskatchewan Research Council, Saskatoon, 2009.

107. NRCan, "Biodiesel Research Shows Positive Results in Saskatchewan," News Release, Natural resources Canada, Ottawa, 13 December 2010.

108. Johnstone, B., "On-Farm Tests Fuel Biodiesel Optimism," *Saskatoon Star-Phoenix*, and "Passing Grades for Fuel," *Regina Leader-Post*, 15 December 2010.

109. Hancock, B., "Partners Announce Plan to Turn Sawdust into Power," *Saskatoon Star-Phoenix*, and "Sawdust to Generate Heat," *Regina Leader-Post*, 3 October 2003.

110. Editorial, "SRC Biodigester Unveiled in Ottawa," *Saskatoon Star-Phoenix*, and "SRC Device Unveiled in Ottawa," *Regina Leader-Post*, 29 April 2014.

111. NRC, "National Research Council and Saskatchewan Research Council Partner to Focus on National Priorities," News Release, National Research Council, Ottawa, 16 July 2015, http://news.gc.ca/web/article-en.do?nid=1001329.

112. Macpherson, A., "NRC and SRC Partner on Fermentation Facility," *Saskatoon Star-Phoenix*, 16 July 2015.

113. Saskatchewan, "Services on Forestry Research and Knowledge to be Reorganized," News Release, Government of Saskatchewan, Regina, 27 May, 2009, http://www.saskatchewan.ca/government/news-and-media/2009/may/27/services-on-forestry-research-and-knowledge-to-be-reorganized.

114. Saskatchewan, "SRC and CLC Announce Strategic Relationship," News Release, Government of Saskatchewan, Regina, 27 January, 2010.

115. SRC, *Real World Solutions. 2011–2012 Annual Report*, Saskatchewan Research Council, Saskatoon, 2012.

116. Saskatchewan, "Action on Energy Conservation," News Release, Government of Saskatchewan, Regina, 16 September, 2002, http://www.saskatchewan.ca/government/news-and-media/2002/september/16/action-on-energy-conservation.

117. Sulatisky, M., Hill, S., Song, Y., Young, K., Gnanam, G., "Neural Control System and Method for Alternatively Fueled Engines," *Canadian Patent* 2,422,970, May 26, 2009.

118. SaskEnergy, "Hydrogen Vehicle Loading and Fuelling Station Launched," News Release, SaskEnergy, Regina, 29 October, 2010, http://www.saskenergy.com/About_SaskEnergy/News/news_releases/2010/NRHydrogenFuellingStation_29Oct2010Final.pdf.

119. Saskatchewan, "Mineral Processing Pilot Plant Backgrounder," Government of Saskatchewan, Regina, 14 September, 2012, http://www.gov.sk.ca/adx/aspx/adxGetMedia.aspx?mediaId=1776&PN=Shared.

120. SRC, "SRC's New Mineral Processing Pilot Plant is Now Open for Business," News Release, Saskatchewan Research Council, Saskatoon, 25 October, 2013, http://www.src.sk.ca/media/pages/news.aspx?itemID=139.

121. PTRC, "2001/2002 Annual Report," Petroleum Technology Research Centre, Regina, 2002.

122. SRC, "Effective EOR Technologies in Tight Bakken Oil Reservoirs: An SRC Consortium R&D Approach," SRC Fact Sheet, Saskatchewan Research Council, Saskatoon, April, 2014, http://www.src.sk.ca/resource%20files/bakken%20oil%20reservoirs.pdf.

123. Yurkowski, M., "Saskatchewan Oil and Gas Outlook," Saskatchewan Ministry of Economy, Regina, April, 2015, http://wbpc.ca/pub/documents/archived-talks/2015/Presentations/Yurkowski%20-%20SK%20Industry%20Outlook%20and%20Updates.pdf.

124. SRC, "Post-CHOPS Oil Reservoirs," SRC Fact Sheet, Saskatchewan Research Council, Saskatoon, September, 2013, http://www.src.sk.ca/resource%20files/post-chops%20oil%20reservoirs%20fact%20sheet.pdf.

125. Schramm, L.L.; Kramers, J.W.; Isaacs, E.E., "Saskatchewan's Place in the Canadian Oil Sands," *J. Can. Petrol. Technol.*, **2010**, *49(11)*, 12-21.

126. Schramm, L.L.; Soveran, D.; Schreiner, B.; Kramers, J.W., "Shale Oil Potential in Saskatchewan," *Cdn. Energy Technol. & Innovation*, **2012**, *1(1)*, 22-28.

127. SRC, 14 – 15 Annual Report, Saskatchewan Research Council, Saskatoon, 2015.

128. Saskatchewan, "SRC Climate Reference Station Celebrates 45 Years and New Equipment," News Release, Government of Saskatchewan, Regina, 28 September 2009, https://www.saskatchewan.ca/government/news-and-media/2009/september/28/src-climate-reference-station-celebrates-45-years-and-new-equipment.

129. Saskatchewan, "Saskatchewan Research Council Unveils New Climate Reference Station," News Release, Government of Saskatchewan, Regina, 31 August 2011, http://publications.gov.sk.ca/details.cfm?p=62217.

130. NASA, "NASA: Climate Change May Bring Big Ecosystem Changes," News Release, NASA Jet Propulsion Laboratory, Pasadena, CA, December 14, 2011, http://www.jpl.nasa.gov/news/news.php?release=2011-387.

131. Bergengren, J.C.; Waliser, D.E.; Yung, Y.L., "Ecological Sensitivity: a Biospheric View of Climate Change," *Climatic Change*, **2011**, *107*, 433-457.

132. Muldoon, J.; Schramm, L.L., "Gunnar Uranium Mine Environmental Remediation – Northern Saskatchewan," Paper ICEM2009-16102, Proc. 12th Internat. Conf. Environmental Remediation and Radioactive Waste Management - ICEM'09/DECOM'09, Liverpool, U.K., October 11-15, 2009.

133. Muldoon, J.; Schramm, L.L., "Gunnar uranium mine remediation project. Northern Saskatchewan," Proc. 33rd Arctic and Marine Oilspill Program (AMOP) Technical Seminar on Environmental Contamination and Response, Halifax, N.S., June 7-9, pp. 383-403, 2010.

134. Schramm, L.L. (2012) Cleaning-Up Abandoned Uranium Mines in Saskatchewan's North, (W. B. Lewis lecture) *Bulletin of the Canadian Nuclear Society*, **33**(2), 17-23.

135. Muldoon, J.; Yankovich, T., and Schramm, L.L., "Gunnar Uranium Mine Environmental Remediation - Northern Saskatchewan," Paper, ICEM2013-96223, *Proc. 15th Internat. Conf. on Environmental Remediation and Radioactive Waste Management, ICEM 2013*, Brussels, Belgium, Sept. 8-12, 2013, 10 pp.

136. Wilson, I.; Allen, D.E.; Schramm, L.L.; Muldoon, J., "Lorado Uranium Mine Environmental Remediation – Northern Saskatchewan," *Proc. 2016 IAEA Internat. Conf. on Advancing the Global Implementation of Decommissioning and Environmental Remediation Programmes*, Madrid, Spain, paper CN-238, 23-27 May 2016.

137. Muldoon, J.; Calette, M.; Schramm, L.L., "Aboriginal and Northern Involvement and Benefits from Gunnar Uranium Mine Environmental Remediation – Northern Saskatchewan," *Proc. IAEA Internat. Conf. on Advancing the Global Implementation of Decommissioning and Environmental Remediation Programmes*, Madrid, Spain, 23-27 May 2016, paper IAEA-CN-238-29.

138. Schramm, L.L., *Gunnar Uranium Mine. Canada's Cold War Ghost Town*, Electronic Ed., Saskatchewan Research Council, Saskatoon, and Amazon.com Inc., 2016.

139. MacPherson, A. "Gunnar Cleanup to Exceed $250M, 10 Times Estimate," *Saskatoon StarPhoenix*, October 17, 2015, Last Updated: February 18, 2016.

140. MacPherson, A. "Overbudget Gunnar Cleanup Federal Responsibility, Sask. Politicians Say," *Saskatoon StarPhoenix*, February 24, 2016.

141. SRC, Annual Report 15 – 16, Saskatchewan Research Council, Saskatoon, 2016.

142. Schramm, L.L.; Nyirfa, W.; Grismer, K.; Kramers, J., "Research and Development Impact Assessment for Innovation-Enabling Organizations," *Canadian Public Administration*, **2011**, *54(4)*, 567-581.

143. Kramers, J., "The Alberta Research Council's Performance Measures," Paper presented at Conference Board of Canada R&D Impact Network Workshop, 11 September, 1998, Ottawa, ON.

144. Kramers, J., "The Alberta Research Council's Performance Measures," Presented at Conference Board of Canada R&D Impact Network Workshop, 5 November, 1999, London, ON.

145. Nyirfa, W.; Schramm, L.L., "Mandate Effectiveness at the Saskatchewan Research Council." *WAITRO News*, World Association of Industrial Technology and Research Organizations, **2005**, *June*, 4-5.

146. Schramm, L.L., "Preliminary Assessment of SRC's Direct Annual Economic Impact on Saskatchewan's General Revenue Fund (GRF)," Confidential internal report, Saskatchewan Research Council, November, 2015.

147. DOE, "Drilling Sideways -- A Review of Horizontal Well Technology and Its Domestic Application," Report, DOE/EIA-TR-0565, U.S. Department of Energy, Washington, April 1993.

148. Jespersen, P.J.; Fontaine, T.J.C., "The Tangleflags North Pilot: A Horizontal Well Steamflood," *J. Can. Petrol. Technol.*, **1993**, *32(5)*, 52-57.

149. Jespersen, P.J., Personal communication to the author at SRC, 2 Sept., 2016.

150. Jespersen, P.J., Personal communication to Brian Kristoff at SRC, 19 Dec., 2015.

151. Graham, A., quoted in "Impact Audit Report: EnCana Resources. Technology: CO_2 Injection for Enhanced Oil Recovery," Impact Audit Report for Saskatchewan Research Council, Spring, 2003.

152. Wilson, M.; Monea, M. (Eds.), *IEA GHG Weyburn CO_2 Monitoring & Storage Project*, Petroleum Technology Research Centre, Regina, 2004.

153. Ralko, J., "Energy-Efficient Houses," *The Encyclopedia of Saskatchewan*, University of Regina, 2006, http://esask.uregina.ca/entry/energy-efficient_houses.html.

154. Paulsen, M., "High-Performance Homes," *Canadian Geographic*, 1 June **2012**, http://www.canadiangeographic.ca/article/high-performance-homes.

155. Ruby, D., "From the Far North: Lessons on How to Slash Fuel Bills," *Popular Science*, **1981**, *October*, 106-108, 144.

156. Holladay, M., "The History of Superinsulated Houses in North America," Presented to British Columbia Building Envelope Council, Vancouver, 22 September 2010, http://www.bcbec.com/docs/5%20-%20History%20of%20Superinsulation%20%5BCompatibility%20Mode%5D.pdf.

157. NRCan, "2012 R-2000 Standard," Building Standard, Natural Resources Canada, Ottawa, 2012.

158. Dumont, R.; Holzkaemper, R., "Moving Towards Sustainability in House Energy Use: Two Saskatchewan Examples, Presented at: *WAITRO Conference*, Saskatoon, August 2006.

159. Dumont, R. "Regina Demonstration Home Predicted to Use 90% Less Energy," *Sask Business*, Apr-May, 2007.

160. Dumont, R., "Demonstration Home Boasts Affordable Energy Savings," *Sask Business*, Jan-Feb, 2008.
161. SRC, "Smart Growth," 2010-2011 Corporate Social Responsibility Report, Saskatchewan Research Council, Saskatoon, 2011.
162. SRC, "Serving Sustainably," 2011-12 Corporate Social Responsibility Report, Saskatchewan Research Council, Saskatoon, 2012.
163. SRC, "Pathway to Sustainably," Corporate Social Responsibility Report 2012-2013, Saskatchewan Research Council, Saskatoon, 2013.
164. SRC, "What Matters," Corporate Social Responsibility Report 2014, Saskatchewan Research Council, Saskatoon, 2014.
165. SRC, "Impacts," SRC Corporate Social Responsibility Report 2015, Saskatchewan Research Council, Saskatoon, 2015.
166. Saskatchewan, "Canada's Biggest Science Experiment a Hit!" News Release, Government of Saskatchewan, Regina, 15 May 2003, http://www.saskatchewan.ca/government/news-and-media/2003/may/15/canadas-biggest-science-experiment-a-hit.
167. Saskatchewan, "Young Saskatchewan Scientists Make a Splash!" News Release, Government of Saskatchewan, Regina, 29 September 2005, http://www.saskatchewan.ca/government/news-and-media/2005/september/29/young-saskatchewan-scientists-make-a-splash.
168. SRC, "Talking Corporate Social Responsibility at SRC," YouTube video, Saskatchewan Research Council, Saskatoon, 24 September, 2012, https://www.youtube.com/watch?v=J8en61bJIA0.
169. SRC, "Take Our Kids to Work Day at SRC," YouTube video, Saskatchewan Research Council, Saskatoon, 8 November, 2016, https://www.youtube.com/watch?v=1LzD57uBy_Y.
170. Calvert, L., Letter to Laurie Schramm at SRC, 4 April 2005.
171. Wall, B., Letter to Laurie Schramm at SRC, 7 December 2015.
172. Historica, "Provincial Organizations Research," Historica Canada, Toronto, 2013, http://www.thecanadianencyclopedia.ca.
173. Western Heritage, "About Western Heritage," Western Heritage, Saskatoon, 2016, http://www.westernheritage.ca/wp39/about/.
174. APRO, "Growing Concerns: Problems and Opportunities of Small and Medium Size Enterprises in Canada," Discussion paper, Association of Provincial Research Organizations, Ottawa, ON, 9 April 1991.
175. I-CAN, "Accessing the System," Innoventures Canada Report 2011," Innoventures Canada Inc., Ottawa, 2011.

LAURIER L. SCHRAMM

www.ingramcontent.com/pod-product-compliance
Lightning Source LLC
Chambersburg PA
CBHW040929030426
42334CB00002B/8